Arctic Studies - A Proxy for Climate Change

Edited by Masaki Kanao,
Yoshihiro Kakinami and Genti Toyokuni

Published in London, United Kingdom

IntechOpen

Supporting open minds since 2005

Arctic Studies - A Proxy for Climate Change
http://dx.doi.org/10.5772/intechopen.73730
Edited by Masaki Kanao, Yoshihiro Kakinami and Genti Toyokuni

Contributors
J. N. Stroh, Sergei Kirillov, Gleb Panteleev, Oceana Francis, Max Yaremchuk, Ekatrina Bloshkina, Nikolay Lebedev, Dimitrios Kostopoulos, Ove T Gudmestad, Efrat Yitzhak, Violetta Gassiy, Natalia Belisheva, Evgeniy Rogozhin, Galina Antonovskaya, Natalia Kapustyan, Irina Basakina, Masaki Kanao

Notice
Statements and opinions expressed in the chapters are these of the individual contributors and not necessarily those of the editors or publisher. No responsibility is accepted for the accuracy of information contained in the published chapters. The publisher assumes no responsibility for any damage or injury to persons or property arising out of the use of any materials, instructions, methods or ideas contained in the book.

First published in London, United Kingdom, 2019 by IntechOpen
IntechOpen is the global imprint of INTECHOPEN LIMITED, registered in England and Wales, registration number: 11086078, 7th floor, 10 Lower Thames Street, London, EC3R 6AF, United Kingdom
Printed in Croatia

British Library Cataloguing-in-Publication Data
A catalogue record for this book is available from the British Library

Additional hard and PDF copies can be obtained from orders@intechopen.com

Arctic Studies - A Proxy for Climate Change
Edited by Masaki Kanao, Yoshihiro Kakinami and Genti Toyokuni
p. cm.
Print ISBN 978-1-78984-099-5
Online ISBN 978-1-78984-100-8
eBook (PDF) ISBN 978-1-83962-795-8

We are IntechOpen,
the world's leading publisher of
Open Access books
Built by scientists, for scientists

4,400+
Open access books available

117,000+
International authors and editors

130M+
Downloads

Our authors are among the

151
Countries delivered to

Top 1%
most cited scientists

12.2%
Contributors from top 500 universities

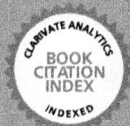

Interested in publishing with us?
Contact book.department@intechopen.com

Numbers displayed above are based on latest data collected.
For more information visit www.intechopen.com

Meet the editors

Dr. Masaki Kanao obtained his PhD from Kyoto University. He is currently working at the National Institute of Polar Research in Tokyo, Japan. He is chiefly interested in Earth's structure and dynamics from geoscience studies. Polar regions, both in the Arctic and Antarctic, have been investigated by using seismological and geophysical approaches. Recently, interdisciplinary studies regarding glacial earthquakes and cryoseismic events in polar regions, particularly around Greenland, have been focusing on involving environmental changes associated with global warming, which is now ongoing in the Arctic. These field-based investigations in polar regions have been contributing to the development of all kinds of Earth sciences from global viewpoints.

Dr. Yoshihiro Kakinami obtained his PhD from Hokkaido University. He is currently working at Hokkaido Information University in Hokkaido, Japan. He is mainly interested in the dynamics of the upper atmosphere. Recently, he has been especially interested in the interaction between solid earth and the atmosphere. Particularly, he investigates the response of the ionosphere and thermosphere triggered by earthquakes, tsunamis, and volcanic eruptions. He is also developing an early warning system for tsunamis by using infrasound emitting from the tsunami source region.

Dr. Genti Toyokuni is an assistant professor at the Department of Geophysics, Tohoku University, Sendai, Japan. He received his BSc (2004), MSc (2006), and DSc (2009) from Kyushu University, Japan. He was a part-time teacher of Earth science at the National Institute of Technology, Kurume College, Japan, during 2007–2009, and a postdoctoral fellow at the National Institute of Polar Research, Japan, during 2009–2011. His research interests include polar seismology (cryoseismology), seismic waveform analysis, numerical modeling of seismic wave propagation, and seismic tomography. He is a member of the Greenland Ice Sheet Monitoring Network (GLISN), and has been participating in the field observation of the Greenland ice sheet every year since 2011 with other US and Japanese GLISN members.

Contents

Preface

The Arctic region is the keystone to understanding the present status of currently ongoing Earth systems and to predicting future images of our planet as viewed from northern high latitudes. The Arctic region, composed of ice-covered Arctic ocean in its center and surrounding fragmentation of the major continents, has been investigated during the last half century through all kinds of scientific studies: bioscience, physical sciences, geoscience, oceanography, and environmental studies, together with the technological domain. This book covers topics on the recent development of all kinds of scientific research in and around the Arctic region, with a view to monitoring the current variations in the extreme environment, affected by remarkable changes in temperature and sea-ice extent, mass loss of ice-sheet and glaciers, and variations in marine and terrestrial ecosystems, including human activities. The most exciting initiative in the Arctic region was the International Polar Year (IPY) in 2007–2008, which was held as part of the 50th anniversary of the International Geophysical Year (1957–1958). The initiative greatly enhanced the exchange of ideas across nations and scientific disciplines to unveil the status of and changes to planet Earth. This kind of interdisciplinary approach helps us understand and address many challenges such as rapid environmental change and its impact on human society. In this regard, this book partially aims to compile achievements of involved projects by the IPY and post era, especially focusing on surface environmental variations associated with climate change such as global warming.

Dr. Masaki Kanao
Associate Professor,
National Institute of Polar Research,
Tokyo, Japan

Chapter 1

Introductory Chapter: Arctic Studies - A Proxy for Climate Change

Masaki Kanao

1. Introduction

The Arctic region is a keystone to understand the present status of the currently ongoing Earth system, such as the climate change, together with to predict future images of our planet as viewed from the northern high latitude. The Arctic region, composed by the ice-covered Arctic Ocean in its center and surrounding fragmentation of major continents, has been rapidly investigated during the last half century

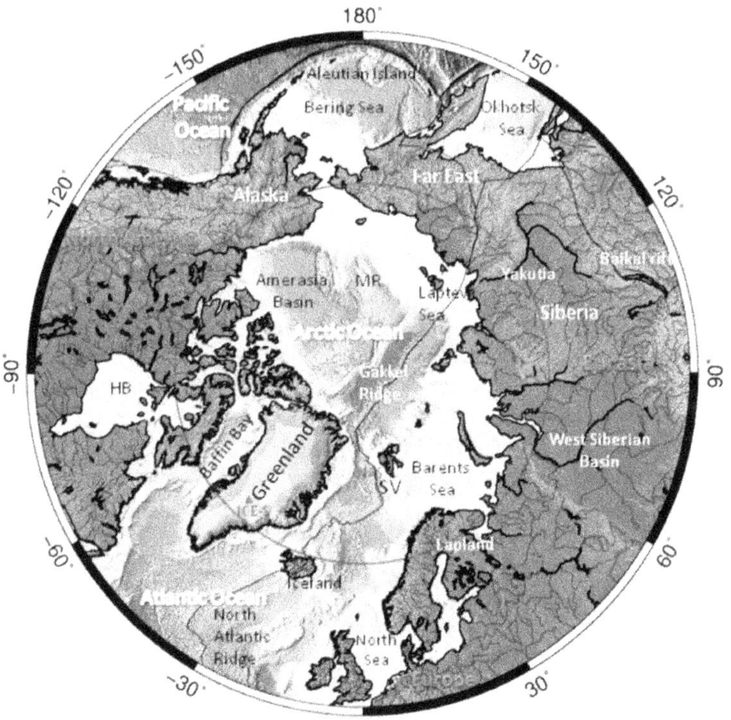

Figure 1.
Surface topography and bathymetry in the Arctic (ETOPO1, [6]) with major geographic location names treated in this review paper. Plate boundaries are after [7]. Red solid circle represents the "Arctic circle" (66.6°N). Abbreviations of representative local names are as follows: SV, Svalbard; MR, Mendeleev Ridge; HB, Hudson Bay; red solid triangles are the permanent stations; KMS, Kamenskoye Station; NAS, Ny-Alesund Station; ICE-S (south). (Original figure prepared for this InTech Book: Arctic Studies).

by all kind of scientific studies including bioscience, physical sciences, geosciences, oceanography, and environmental studies, together with the technological domain. Major location names in the Arctic region are illustrated in **Figure 1**.

2. Scope of the book

This book covers the topics on recent development of all kinds of scientific researches in and around the Arctic region, with a viewpoint to monitor the current variations in the extreme environment, affected by remarkable changes in temperature and sea-ice extent, mass loss of ice-sheet and glaciers, and variations in marine and terrestrial ecosystem including human activities. Multidisciplinary and interdisciplinary approaches are beneficial in polar researches, focusing on the interrelated perspectives which will be needed to understand and quantify their connections with the prediction of future climate system change.

In this concern, particularly the cryosphere system is likely to be influenced by temporal–spatial variations in surface environmental conditions in the Arctic, and continuous researches of their variability provide direct evidence of climate change. For instance, large glacial-involved earthquakes are the most prominent phenomena found particularly around Greenland [1–3]; the new innovative studies conducted by glaciological and geophysical investigations are strikingly encouraged by long-term monitoring observations under extreme conditions in the area.

The most exciting initiative in the Arctic region was the International Polar Year (IPY) in 2007–2008, which had been conducted at the 50th anniversary of the International Geophysical Year (IGY 1957–1958). The IPY was a big international program composed of multidisciplinary science branches: upper atmosphere, meteorology, glaciology, geosciences, oceanography, and biosciences conducted by a significant number of polar scientists involved [4]. The initiative significantly enhanced the exchange of ideas across nations and scientific disciplines to unveil the status and changes of planet Earth. This kind of interdisciplinary approach helps us understand and address grand challenges such as rapid environmental change and its impact on human society.

The recent seismological research achievements in the polar region, for instance, are compiled in the special issue on "Polar Science" [5]. By taking into account the above concerns, however, this book partially aims to collect many discipline achievements by a significant number of involved projects at the IPY and post era, primarily focusing on surface environmental variations associated with climate change such as global warming.

It is also mentioned that, moreover, the contents in this book intend to appeal not only to the polar scientists but also to all general public who are interested in the present status of the Arctic region. It is hopeful that this book could provide remarkable knowledge and new understanding regarding current environmental variations and the past Earth's history within the global dynamics. The book could surely attain fruitful information on an advance of frontier researches in the Arctic region which are currently suffering significant effects by climate change within the global system.

Author details

Masaki Kanao
National Institute of Polar Research (NIPR), Organization of Information and
Systems (ROIS), Midori-cho, Tachikawa-shi, Tokyo, Japan

*Address all correspondence to: kanao@nipr.ac.jp

IntechOpen

References

[1] Ekström G, Nettles M, Tsai VC. Seasonality and increasing frequency of Greenland glacial earthquakes. Science. 2006;**311**:1756-1758

[2] Nettles M, Ekström G. Glacial earthquakes in Greenland and Antarctica. Annual Review of Earth and Planetary Sciences. 2010;**38**:467-491

[3] Clinton JF, Nettles M, Walter F, Anderson K, Dahl-Jensen T, Giardini D, et al. Real-time geophysical data enhance Earth system monitoring in Greenland. Eos, Transactions American Geophysical Union. 2014;**95**:13-24

[4] Krupnik I, Allison I, Bell R, Cutler P, Hik D, Lopez-Martinez J, et al. Understanding Earth's Polar Challenges: International Polar Year 2007-2008— Summary by the IPY Joint Committee. Edmonton, Alberta: Art Design Printing Inc; 2011. pp. 457-476

[5] Kanao M, Zhao D, Wiens DA, Stutzmann E. Recent advance in polar seismology: Global impact of the international polar year–overview. Polar Science. 2015;**9**:1-4. DOI: 10.1016/j.polar.2014.12.003

[6] Amante C, Eakins BW. ETOPO1 1 Arc-Minute Global Relief Model: Procedures, Data Sources and Analysis. NOAA Technical Memorandum NESDIS NGDC-24, National Geophysical Data Center, NOAA; 2009. DOI: 10.7289/V5C8276M

[7] Bird P. An updated digital model of plate boundaries. Geochemistry, Geophysics, Geosystems. 2003;**4**:1027. DOI: 10.1029/2001GC000252

Changes in Arctic Ocean Climate Evinced through Analysis of IPY 2007–2008 Oceanographic Observations

J.N. Stroh, S. Kirillov, G. Panteleev, O. Francis,
M. Yaremchuk, E. Bloshkina and N. Lebedev

Abstract

Full-depth hydrographical surveys conducted in 2007–2009 during the International Polar Year (IPY) collaboration provide an accurate snapshot of the Arctic Ocean (AO) hydrography at a time when the Arctic Ocean Oscillation (AOO) index was highest in recent record. We construct pan-Arctic temperature and salinity (T/S) reference states from these data using variational optimal interpolation and discuss some key differences between the 2007–2009 state and a similarly constructed climatology from historical 1950–1994 Russian archives. These data provide a recent, known reference state for both qualitative and quantitative future AO climate change studies. Furthermore, we present an analysis of sea-surface height (SSH) and upper-layer circulation constructed from the IPY data via 4DVar data assimilation and use them to examine circulation and freshwater source changes visible during IPY.

Keywords: Arctic Ocean climate, Arctic Ocean observations, data assimilation, gridded IPY climatology

1. Introduction

During the International Polar Year (IPY) 2007–2008, the international scientific community completed an intensive physical survey of the Arctic Ocean (AO). Many countries and institutions contributed to this effort, which generated a significant number of in situ hydrographical observations including stationary full-depth profiles of temperature/salinity (T/S) from conductivity-temperature-depth instruments (CTD) and partial-depth profiles of the upper ~700 m along Lagrangian tracks followed by Ice-Tethered Profiler (ITP) affixed to sea ice, measurements of T/S along the tracks followed by submarine gliders near coastal areas, and a small number of profiles from less accurate expendable CTD and expendable bathythermograph (XBT) instruments.

Arctic T/S distribution is governed largely by water inflow and outflow through the major gateways, the properties of those waters, and regional circulation. AO sources include the warm saline waters advected with the Norwegian current from

North Atlantic [1], the fresh (relative to mean AO salinity) Pacific waters (PW) entering through Bering Strait [2], and the summertime fresh and warm river discharge from the Siberian and North American rivers [3, 4]. AO export occurs primarily through Fram Strait by way of the Transpolar and East Greenland currents and also through less-studied Canadian Archipelago transport [5]. Within the AO, shelf-basin exchanges are typically restricted to bathymetric features, such as Herald and Barrow Canyons in the Chukchi Sea [6] and St. Anna Trough north of Severnaya Zemlya in the Kara Sea [7]. Regional circulation is governed by topographically steered boundary currents along shelf breaks and other topographic features, restricted by density fronts between water masses of disparate origin, and subjected to external forcing including surface heating (cooling) in summer (winter) with significant effects from sea-ice melt/freeze and large-scale atmospheric pressure systems [1, 8–10].

The IPY effort occurred at a significant time, coincidental with the largest recent positive Arctic Ocean Oscillation (AOO) index, a measure derived. From the central AO sea-surface height gradient over the central AO, which indicates strength of large-scale anticyclonic flow [11]. Prior to IPY, the AOO index had been in an overall positive regime for nearly two decades, while historical records suggest a sub-decadal frequency [9] (updated at www.whoi.edu/page/preview.do? pid=66578). Other modes of regional oscillation occur with timescales of 60– 80 years [12]. At the same time, in summer 2007, winds associated with the Beaufort High remained predominantly anticyclonic—a feature common to the Arctic winter but unusual for summer [13]—so Beaufort Gyre (BG) sea-level response to atmospheric forcing strengthened the AOO. Additionally, 2007 was a monumental year for river discharge; Eurasian river discharge surpassed the 2002 record by nearly 10% [14]. The effects of these drivers, whether purely anomalous or the result of long-term variability, relate to pronounced recent changes in the Arctic marine climate system and were witnessed by the IPY survey efforts.

Over a decade has passed since many of the observed strong and rapid warming trends were confirmed as both present and underway. Older climatologies may be inadequate for the study of more recent changes as they may depend much on pre-1996 data when the positive AOO regime was not such a strong and permanent feature of the region, thicker and more extensive sea ice regulated mechanical and thermodynamical fluxes with the atmosphere, continental riverine discharge was less, and the Lomonosov Ridge roughly defined a partition of the Arctic between Atlantic and Pacific upper-ocean layers. The thermal state of the AO over the past decade is above the long-term average, and this warming greatly affects both hydrographic and ice-related processes observed in the high latitudes; changes occurring under these new conditions are of particular interest [5, 13, 15–23].

Unlike other oceans, vertical stratification of AO water masses is governed more by salinity than temperature, and density gradients readily allow for decomposition of profiles into differentiated, typically noninteractive layers. Away from regions of significant freshwater influence, the vertical distribution of water masses throughout the AO generally comprises a well-mixed surface layer occupying the upper 50–100 m, underlain by a layer of intermediate water of Pacific origin (absent in the eastern, Atlantic domain of the Arctic), followed by a layer of warmer more saline water of Atlantic origins and an Arctic deepwater bottom layer. Importantly, the halocline of Pacific-originated waters overlies the warmer Atlantic water in the ocean in the Pacific sector [24], buffering sea ice from the warmer Atlantic waters below. The presence, thickness, and specific properties of each layer vary laterally throughout the Arctic [10, 25], and one may distinguish between layers of waters of Pacific and Atlantic origin on the depth of local haloclines and isotherms characteristic within each column.

The remote nature of the AO, together with practical difficulties in observation and navigation due to sea ice and sparse infrastructure, makes in situ sampling of the AO expensive and occasional. Satellite monitoring of the ocean surface is possible but inhibited by ice cover and clouds. Unfortunately, the accuracy of the satellite-surface observations and their processed (i.e., L2–L4) products is often far from optimal: they may contain large errors due to poor calibration, mask large portions of the AO for sea ice and thus lack of coverage over the central AO, and may contain anachronistic assumptions in their post-processing algorithms [26]. Modeling efforts and other interdisciplinary studies in need of static background ocean data may need to rely on gridded products that are biased toward older AO regimes or large amounts of surface observations from satellite. Further, climatological studies using older reference states for trend analysis may suffer from amplified trend errors. For example, the Arctic portion of the most recently available Polar Science Center Hydrographic Climatology (PHC 3.0, updated from [27]) is based on historic observations through 1993 [28, dataset g01961].

The concerns listed above motivate this work, which presents a 2007–2009 AO stationary analysis state inferred from algorithmic data conditioning of pan-Arctic hydrographical surveys and other at-depth observations to provide a snapshot of the non-coastal ocean state with an emphasis on the intermediate layers. The result is a dataset of gridded T/S available in NetCDF at http://bit.ly/2M6qsJ9, from which this chapter discusses mapped water masses and their differences relative to those mapped from 1950 to 1994 climatology. We also use 4DVar data assimilation to establish an analysis of major circulation changes during IPY relative to the climatological mean and discuss the evident anomalies of July–December 2008 [29]. The remainder of this chapter is organized as follows: Section 2 discusses the in situ data and the production algorithm for the gridded fields, Section 3 presents an atlas of water mass properties for the IPY and their differences from historical data fields, Section 4 discusses changes in the AO water mass distribution and thermal state evident from the use of IPY data and derived climatology, Section 5 presents analysis of circulation anomalies during the IPY, and Section 6 concludes the chapter.

2. Observational data and gridding

As part of an IPY initiative, approximately 13,000 CTD/xCTD/XBT profiles along with ITP data were curated into a central database of AO T/S observations from contributors in Japan, Norway, Russia, Canada, the USA, Germany, Poland, Sweden, and China. Stroh et al. [[26], Figure 1] show the location of profiles over the AO, of which only the IPY CTD and ITP data during 2007–2008 are used here. CTD observations during the sea-ice minimum months of August–October account for approximately 40% of all ship-borne profiles, while wintertime November–March account for approximately 30%. ITP apparatuses provide a more temporally uniform stream of profile data for the uppermost ~700 m throughout the year; ITP data were collected and made available by the Ice-Tethered Profiler Program [30, 31] based at the Woods Hole Oceanographic Institution (http://www.whoi.edu/itp).

The Data-Interpolating Variational Analysis tool (DIVA, [32]) is a robust finite element-based optimization tool for gridding large 2D, 3D, and 4D datasets and includes error estimates of the analysis. This freely available program, developed by the GeoHydrodynamics and Environment Research, was applied to the observational data described above to construct static full-depth fields on an equal-area polar-centered grid with 50 km resolution. Interpolation to 51 vertical levels occurs level-wise within DIVA, to which an internally applied stability algorithm ensures that analyses remain hydrodynamically stable with respect to density throughout the gridding. Bathymetric masking was inferred from the International Bathymetric Chart of the Arctic Ocean [33], and regions with depth less than 200 m are masked. The correlation length scales for observations correspond to three grid cells with a signal-to-noise ratio of 10%. The same procedure applied to historical observations collected during 1950–1994 (privately archived at the Arctic and Antarctic Research Institute of Russia) generates mean climate dataset for that period, which is used to contrast the gridded IPY data.

3. Water mass distribution maps

From the gridded T/S analyses for the 1950–1994 and IPY periods, water mass properties reveal qualitative differences between them. The use of density-related properties to distinguish water masses is less certain than chemical analysis [22, 34]. Scarcity of widespread chemical tracer surveys precludes such an approach here, and analysis based on the more common T/S data is adopted. This work chooses to map Atlantic water (AW) and summer Pacific water (SPW) for both their simplicity of definition and importance in the freshwater (FW) and thermal budget of the AO. Characteristics used to identify AW and SPW are adapted from [25] and [35, 36], respectively, and are described below.

The AW distinguishes an intermediate layer of warm water of Atlantic origin that has entered the Arctic Basin through deep coastal channels and bathymetric steering. Over-basin AW typically has S ≥ 34.8 PSU with T ≥ 0°C despite heat loss along the Eurasian shelf. SPW denotes relatively fresh waters with 31 PSU ≤ S ≤ 33 PSU and T ≥ −1.4°C entering the AO through the Bering Strait which have cooled after residence on the shallow Chukchi Shelf and include substantial meteoric FW [21, 35]. These low-density waters form a subsurface layer in the western Arctic typically at depths between 50 and 100 m and often include a local temperature maximum [37, 38].

In **Figures 1–11**, left-side plots show the identified field for the IPY dataset, while the right-side plot shows the corresponding anomaly field relative to the

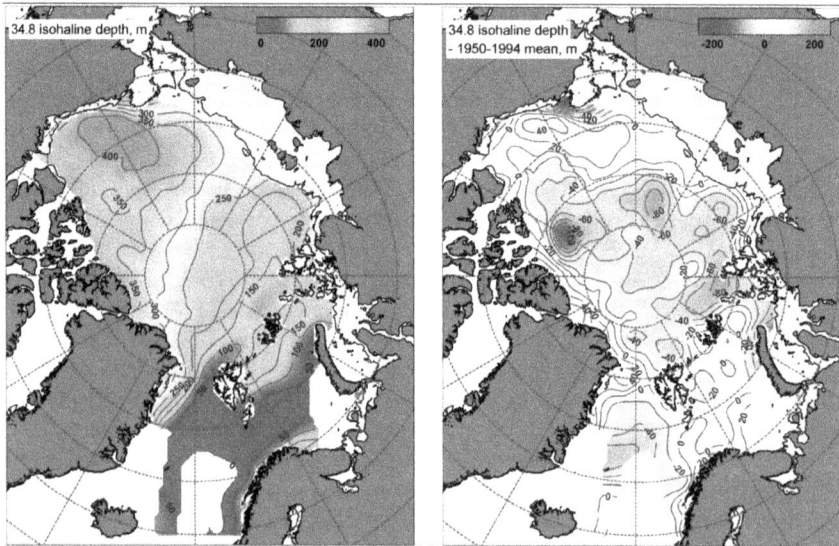

Figure 1.
34.8 PSU isohaline depth. IPY (l) and anomaly (r).

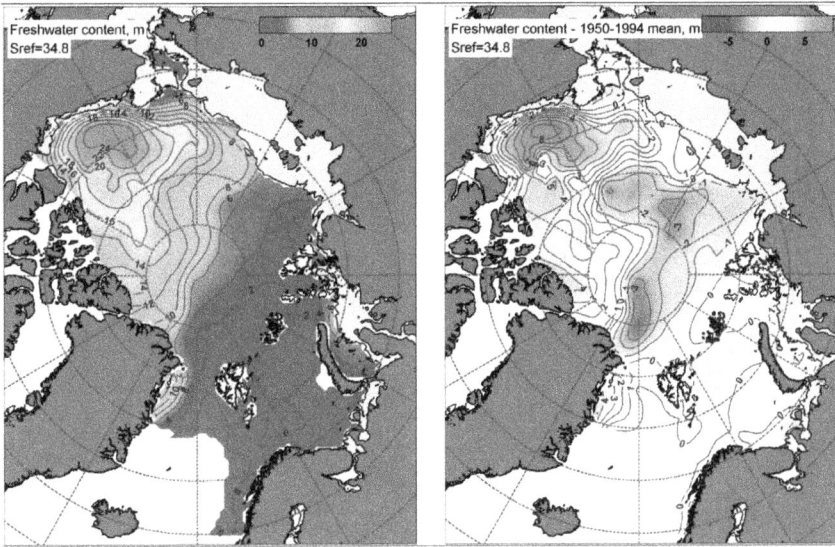

Figure 2.
FWC relative to 34.8 PSU isohaline. IPY (l) and anomaly (r).

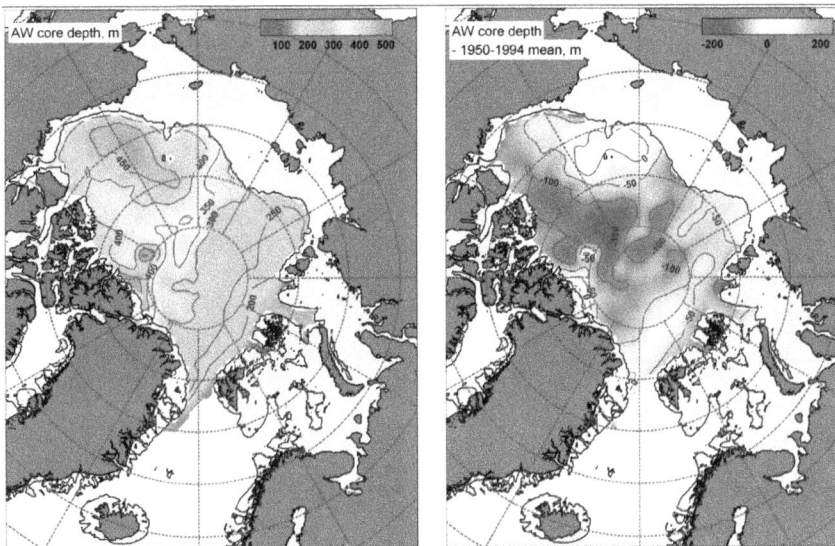

Figure 3.
AW core depth. IPY (l) and anomaly (r).

Russian 1950–1994 archive. We refer to each such pair singularly as a figure and distinguish between the field and its anomaly in context. **Figure 1** maps the 34.8 PSU isohaline depth. **Figure 2** shows the integrated FW content (FWC), in meters of freshwater, with respect to 34.8 PSU.

Figures 3–7 plot the AW core depth, core temperature, heat content, lower boundary depth, and upper boundary depth, respectively. AW here is defined as waters composing a continuous vertical region of positive temperature bounded by 0°C isotherms, which define herein the lower and upper AW boundary depths. The AW core depth and temperature are adopted to be the depth and value of the

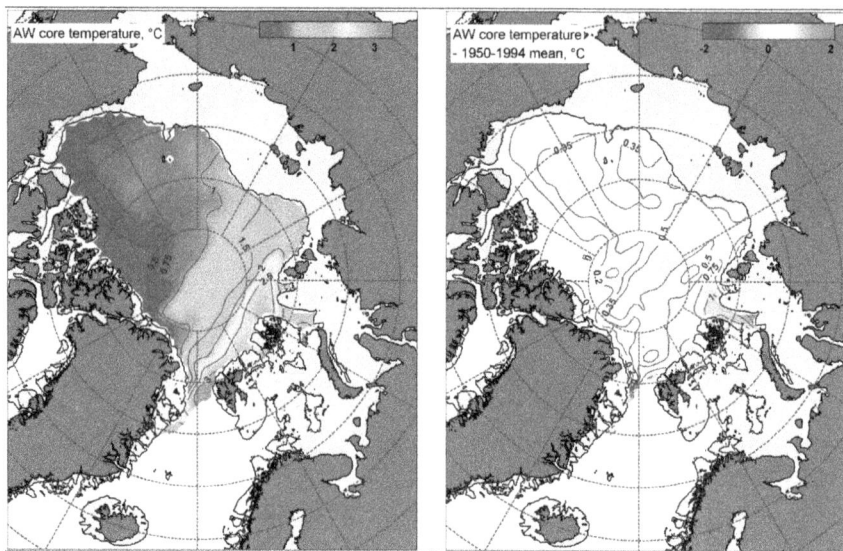

Figure 4.
AW core temperature. IPY (l) and anomaly (r).

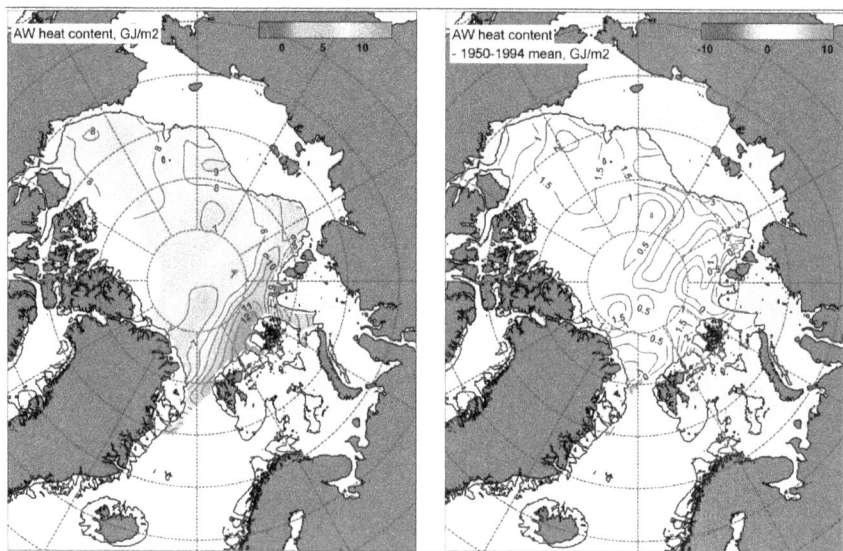

Figure 5.
AW heat content. IPY (l) and anomaly (r).

temperature maximum within the AW layer. Total heat content is calculated as the vertical integral of specific heat with respect to −1.8°C between AW boundaries.

Insufficient deep data in near the Canadian Archipelago precludes a resolution of the AW lower boundary and consequently of the heat content in that area.

Figures 8–11 show calculated fields for summer Pacific water, which exists only on the Pacific side of the Arctic. SPW is defined by a local temperature maximum

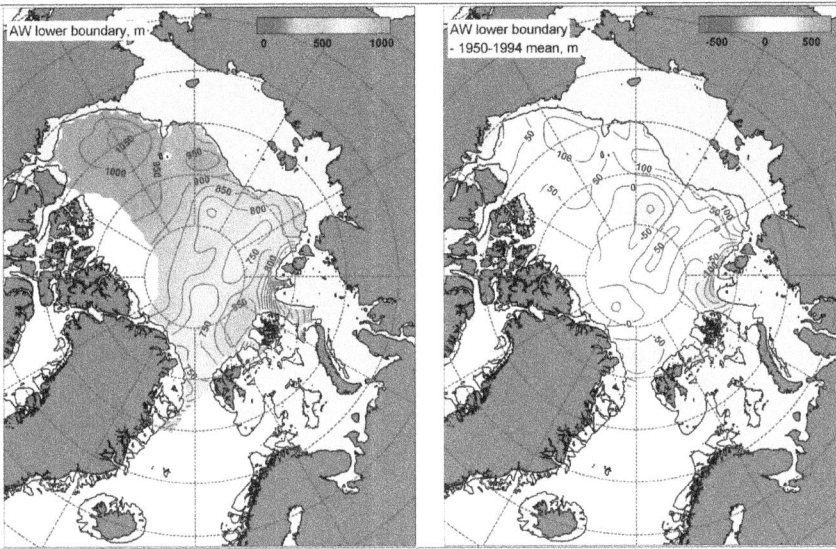

Figure 6.
AW lower boundary. IPY (l) and anomaly (r).

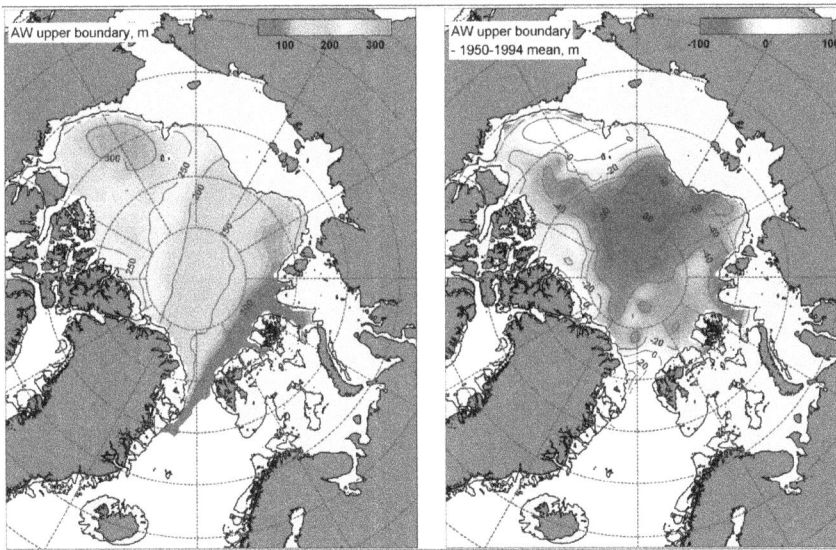

Figure 7.
AW upper boundary. IPY (l) and anomaly (r).

occurring below the surface mixed layer within the salinity range 30.5–33.0 PSU
[35]. Upper and lower SPW boundary depths are determined T ≥ −1.4°C and
salinity restriction to that range. **Figure 8** maps the depth of the maximum temperature found in SPW, and **Figure 9** identifies these maxima. **Figures 10** and **11** show
the lower and upper boundary depths of SPW.

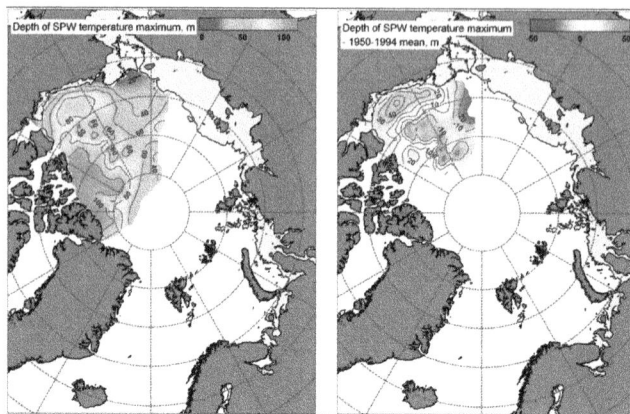

Figure 8.
Summer PW depth of T$_{max}$*. IPY (l) and anomaly (r).*

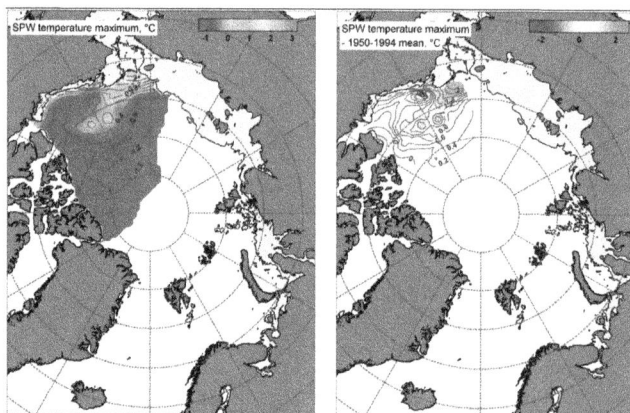

Figure 9.
Summer PW T$_{max}$*. IPY (l) and anomaly (r).*

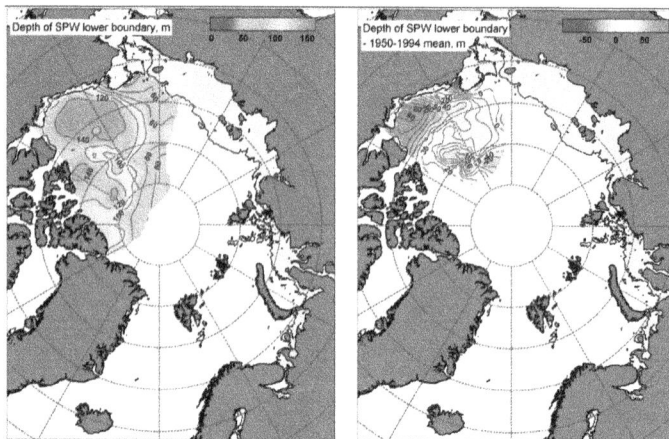

Figure 10.
Summer PW lower boundary depth. IPY (l) and anomaly (r).

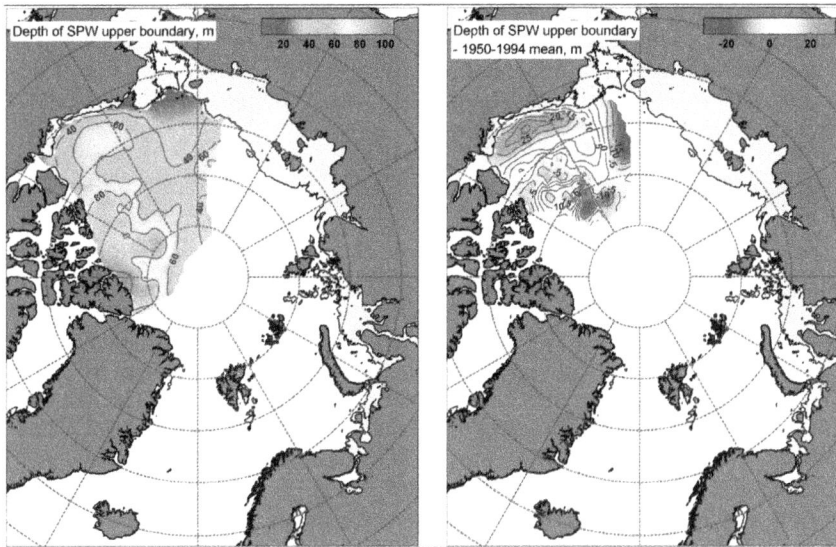

Figure 11.
Summer PW upper boundary depth. IPY (l) and anomaly (r).

4. Changes inferred from T/S observations

In general, the vertical and spatial patterns of hydrographic parameters in the AO and adjacent North Atlantic had undergone considerable changes by IPY although the large-scale distributions of the water masses align with the historic climatology. Readers unfamiliar with AO geography and its bathymetric features are encouraged to follow this discussion with an atlas, e.g. https://geology.com/articles/arctic-ocean-features/.

4.1 Atlantic waters

Elevated pan-Arctic heat content due to the extraordinary heat transported to the AO from the North Atlantic is a significant change evident during the IPY period. Advection of relatively warmer AW resulted in anomalous hydrographic state formation over the entire deep Arctic Basin [17, 38]. The temperatures within the core of AW were observed 0.3–1.0°C higher than climatic values; mean changes are ~0.65°C over the Eurasian Basin and ~0.25°C over Canada and Makarov basins.

Of further note is the warm tongue of AW that appears to be topographically steered by the Lomonosov Ridge; **Figure 4** shows a clear 0.5°C core temperature anomalous increase extending from the Laptev Sea toward the Greenland Shelf. This feature resides at a depth of about 275 m, ~75 m surfaceward of the historic AW core depth per **Figure 3**. Over the Makarov Basin, AW expanded ~50 m deep into the column [39], while the AW core depth has moved 100–150 m surfaceward with an associated 0.5–1.0 GJ/m^2 increase in associated heat content. Similar changes including the AW moving surfaceward and retaining more heat at depth are present throughout most of the AO indicating stronger potential influence on ice-related processes [40].

By 2007, the intermediate AW layer had deepened and thickened in the Pacific sector [23], but the changes are heterogeneous over the central and Eurasian basins.

In particular, the net AW layer thickness appears to have thinned over the Amundsen Basin, which is likely a mass-balance response to the thickened layer observed on the Pacific side of Lomonosov Ridge. Within the western side of Fram Strait, the AW layer has thickened by roughly 70 m, moving 20 m closer to the surface without change in the core depth.

4.2 Pacific water

Figure 2 shows another of the most drastic changes in the Arctic—the change in freshwater distribution. As a proxy for the AW-PW upper-ocean front in the central Arctic, the strong FW anomaly gradient illustrates the change from the Lomonosov Ridge to the Alpha-Mendeleev Ridge (AMR) system [22 and references therein, 41]. Further, the boundary marking the extent of present SPW in **Figures 9–11** tracks very directly the local bathymetric minimum of the AMR. Estimates shortly after IPY show that FWC in the Eurasian domain decreased by nearly one-quarter, while the American domain increased by the same percentage [16, 42]. The influx of PW through Bering Strait was near a record high in 2007, importing anomalously large FW volume and thermal input [20].

The loss of FWC near the pole and in the western sector likely results from cyclonic AO moving more AW toward the eastern Amerasian Basin. Simultaneously, the wind-forced anticyclonic BG stored fresher SPW in the Pacific sector, accumulating an average of 4 m FWC on the Pacific side of the front. Much of this FW had been in place prior to 2007; the IPY FWC in the Beaufort Sea is nearly identical to that found for 2006 [21]. Carmack et al. also find that sea-ice freeze/melt accounts for a net loss of FWC in the Beaufort Region, with riverine water and PW contributing roughly half of the regional FW [21]. Ge et al. find that the mean annual Yukon River outflow, the most significant meteoric source included in SPW, increased 8% between 1977 and 2006 [43].

An increasing trend in Eurasian catchment outflow also is evident [14] and related to changes in permafrost [44] and temporal changes in continental hydrological cycles [45]. Increased Siberian runoff suggests the apparent decreases in FW volumes adjacent to the Laptev and East Siberian seas arising from changes in seasonal ice and the regional dominance of AW, but these source changes alone do not explain FW accumulation observed in the Beaufort Sea during IPY and beyond [46]. Data-conditioned modeling of the 2008 circulation [29] suggests that this accumulation may be supported by transport from the Lincoln Sea [47] and/or regions north of Greenland.

Changes in the organization of water masses have also affected the outflow of AO through Fram Strait, located between Greenland and Svalbard. The Transpolar Drift mode arising from the cyclonic AOO regime impedes PW from reaching the continental shelf north of Greenland. Consequently PW may only exit the AO via the Canadian Archipelago [19], which has been shown to be a significant but variable route for AO export [5, 48, 49].

4.3 Directly observed from ITP data

The gridded IPY data do not resolve a surface layer. Sea-surface temperature and salinity (SST and SSS, respectively) are temporally variable as they depend on the strongly seasonal Arctic diurnal effects. Additionally SST/S in the AO depends seasonally on sea-ice-related processes such as meltwater strata, brine rejection, rapid wintertime heat loss through sea-ice leads, etc. Models and SST satellite data products often assume a surface freezing temperature (FT) of $-1.8°C$, which assumes background salinity of ~32.86 PSU. At that T/S state, FT sensitivity is

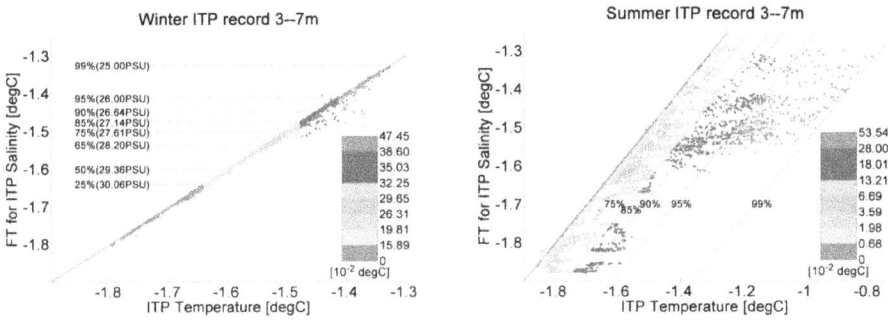

Figure 12.
Shallow ITP-observed temperature. Winter (l) and summer (r).

~0.1°C per −0.01 PSU so that inaccuracies in background salinity amplify errors in associated freezing temperature.

Figure 12 illustrates the inaccuracies of these assumptions by examining the relationship between near-surface temperatures observed by 2006–2009 ITP and FT calculated from the associated salinity. Observations are primarily over the Pacific sector and central Arctic. The thick diagonal line shows exact correspondence between observed T and FT. Colors indicate binned values of T + 1.8°C (T-FT) in winter (summer) in the left (right) plot, with dashed lines demarcating percentiles as labeled. In winter months of November–April, all observations correspond to freezing point, but only about 25% of measurements have T ≤ −1.64°C, the freezing temperature associated with ~30 PSU. In summer months of May–October, temperatures clearly depart from freezing, but only ~25% of measurements differ from freezing by more than 0.05°C. In both summer and winter, the vertical structure of the plots demonstrates inaccuracy of the −1.8°C at ~32.86 PSU assumption; surface waters in the western Arctic have salinities in the range 30–32 PSU.

5. Changes in circulation

5.1 Quasi-stationary "climatological" circulation

Freshwater changes throughout the Arctic relate to changes in geostrophic current distributions. Over basins, the strengthened FW gradient between the Pacific and Atlantic sectors led to a very significant sea-surface height (SSH) changes, which in turn gives rise to changes in geostrophic currents [16]. The strengthening of geostrophic currents in the Pacific sector is suspected among the factors for the reduction of multiyear ice over the Canadian Basin [50]. Other factors include deepening AW over the Canada Basin since 2004, enhancing the strength of the BG, and its accumulation of freshwater [23]. A recent study demonstrates that atmospheric modulation of geostrophic boundary currents and SSH quantifiably relates to the Northern Hemisphere annular mode strength [51].

To analyze the quantitative difference in the mean circulation during the IPY period with respect to the climatological circulation, the IPY dataset was conditioned using the four-dimensional variational (4DVar) data assimilation (DA) approach [52, 53] in two ways. To find a quasi-stationary solution, the process uses 4DVar optimization of an ocean model forced by the corresponding heat, salt, and momentum fluxes inferred from NCEP/NCAR reanalysis and regional Pan-Arctic Ice-Ocean Modeling and Assimilation System (PIOMAS). In the nonstationary reconstructions, all available T/S data were averaged for model grid bins, and these

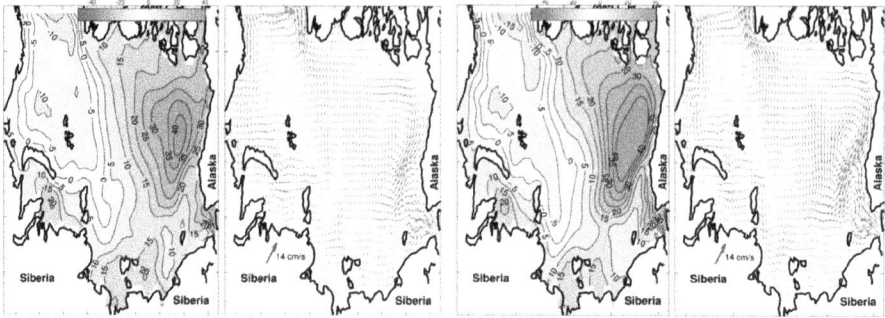

Figure 13.
Quasi-stationary model reconstruction of SSH and near-surface currents from historical 1900–2006 data (left plots) and the IPY data (right plots).

meaned observations were assimilated through the conventional 4DVar DA approach using a semi-implicit ocean model (SIOM) with resolution of 65 km; a framework of the algorithm is described in [52, 54].

The resulting quasi-stationary SSH maps and near-surface currents are shown in **Figure 13**. A comparison indicates the essential reorganization of the circulation in the AO evident during IPY. The most notable feature is the strong intensification and shift of the BG toward the Alaska. IPY SSH patterns are characterized by a pronounced BG dome which attains a central height greater than 50 cm, while the typical climatological SSH is only about 40 cm. This difference results from intensified westward flow along the Alaskan and Chukchi Sea continental slope. There is also a clear re-centering of the BG resulting from the shift of the Transpolar Drift axis toward the Canada Basin; this agrees well with the recent analysis of the freshwater content and circulations conducted by [55].

5.2 Anomalous 2008 circulation

The application of the more advanced 4DVar reconstruction of nonstationary circulation for July–December 2008 indicates stronger circulation than those directly detected from the in situ IPY dataset.

The SIOM-4DVar reconstructed bimonthly evolution of SSH and circulation at 250 m during July–December 2008 is shown in **Figure 14**. The SSH patterns are characterized by a pronounced BG dome which gets slightly stronger in November–December (**Figure 14**, right) attaining a 40 cm central elevation. Compared to the relatively smooth and symmetric SSH derived through optimal interpolation of observations (e.g., [16]), the DA-reconstructed SSH reveals finer features consistent with the observations. During September–October, the SSH pattern is characterized by a secondary SSH maximum at 74°N 140°W, which tends to erode by the end of the year but still persists as a tongue spreading toward Alaska along 140°W. This feature is seen in the AVISO anomalies averaged over the second half of 2008 [29].

Another prominent feature is a zonally spreading trough in the region between 72°N and 80°N from Severnaya Zemlya to the Bering Strait. The emergence of this depression could be one of the causes of intensification of the Bering Strait transport due to the increase of the large-scale sea level difference between the Chukchi and Bering Seas. This is supported by the analysis of Woodgate et al., who estimated the force balance controlling the flow through the Bering Strait and found a significant increase of the pressure head in 2007–2011 with respect to the 1997–2006 period when the Bering Strait transport was smaller [20 and Figure 1h therein]. The behavior of the SSH lowering is shown in the bimonthly SSH fields averaged over

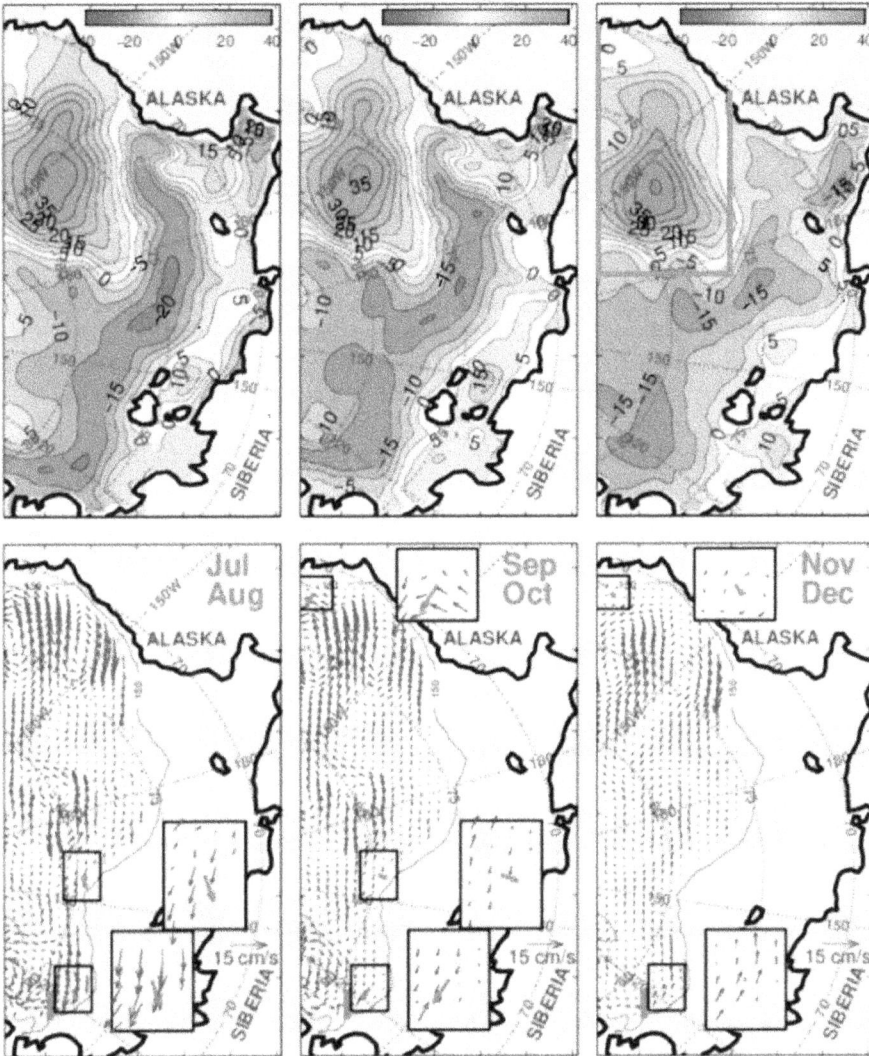

Figure 14.
Bimonthly averaged fields for SSH in cm (upper panels) and velocity at 250 m depth in cm/s (lower panels). NABOS mooring-observed velocities are shown by insets. The boundaries used for calculating the total FW flux around BG are shown by the thick gray line in the top-right panel; the gray line adjacent to the figure panel frame is the "eastern" boundary around BG through which FW transport is positive (gyreward).

the area north of the Bering Strait (upper panels in **Figure 14**), the heights of which are estimated to be −11, −10, and −6 cm, respectively. This is consistent with the seasonal decline of the Bering Strait inflow from 1.1 Sv in July–August to 0.5 Sv in November–December 2008 [20].

The effect of the abovementioned SSH decrease on the transport pattern in the region of the AW inflow is of particular interest. During July–August 2008, the negative SSH anomaly is closely attached to the coastline, creating a positive cross-shelf SSH gradient and a westward geostrophic transport of −2.9 Sv along the shelf break (lower-left panels in **Figure 14**). The effect becomes less visible by the end of the year as the negative SSH anomaly detaches from the continental slope; the total transport relaxes to eastward values of 0.8 and 1.0 Sv, respectively, for the September–October and November–December periods. This identified flow reversal agrees well with moored velocity observations from the Nansen and Amundsen Basins Observational System (NABOS, http://nabos.iarc.uaf.edu/data), which are indicated by red arrows in **Figure 14** but were not used to obtain the optimized solution.

The DA results immediately provide us with quantitative FWC estimates and permit identification of the regional FW. In particular, the total FWC within the volume bounded within [70.25, 80]°N × [140, 170]°W above 400 m depth was found to be about 20,700 km^3, which is slightly less (~5%) than that found in literature [update from 46]. A possible source of this difference is a smaller area of the integration for the 4DVar solution and the offshore displacement of the BG observed in 2008.

To assess the FW origin accumulated FWC in the BG, FW transports across the eastern, southern, and western boundaries were estimated 0.08, −0.005, and −0.075 Sv, respectively (positive-oriented gyreward); the boundaries are shown in the top-right panel of **Figure 14**, where the eastern boundary abuts the figure boundary and the southern one intersects the Alaska coast. Calculated transports suggest that observed changes in the BG FWC were generally caused by the FW transport changes confined to the latitude band of 72–77°N at the eastern boundary of the model domain.

6. Summary

This work introduces an IPY snapshot ocean climatology and discusses freshwater and thermal changes in two principle water masses to establish, in perspective, subsurface changes over the central AO as well as consequences of surface freshening. It focuses only on the ocean and readily neglected continental shelves where important water mass-forming processes occur [56] but enhanced mixing impedes analysis based on T/S, any resolvable changes in Arctic Bottom Water, and a direct analysis of sea ice which requires an extensive discussion of the atmosphere and its variability [57] which are beyond the scope of this presentation.

Changes in the AO are not monotonic as they result from cyclic and quasi-cyclic changes in various superimposed feedback-entangled geophysical components in addition to trends in their background values. Changes may arrive in short bursts or "pulses" and may undergo periods of relaxation toward long-term means. The intensive pan-Arctic IPY survey provides evidence of an AO undergoing significant changes and departure from the longer-term mean of the late twentieth century— responding to variations in source content (from the Atlantic, Pacific, and continental waters) and the resulting changes in freshwater and heat distribution; atmospheric forcing, induced SSH gradients, and their associated geostrophic responses; and relative volume and means of exit of various water masses present in the AO. During IPY, many of these components appeared to be establishing new records. In the decade following, 2011–2012 set records for associated components such as river outflow, Bering Strait inflow, sea-ice minimum, and Arctic cyclone strength—some of which may have been surpassed those of 2016–2017. From this perspective, conditions of the AO during IPY 2007–2008 show that the region is in transition toward a "new normal," and a gridded IPY dataset provides a useful reference state for establishing how far that transition has progressed.

A model-DA system was also applied and may quantify the observed difference in the T/S distribution bought on climatological and seasonal temporal scales. The reconstructed mean 2007–2009 AO circulation clearly identified global shifts in the BG and axis of the transpolar drift. Both results are consistent with other qualitative analyses. Analysis of the reconstructed nonstationary circulation for July–December 2008 allowed quantification of several anomalous circulation features including:

a. A reversal of the total transport in the AW inflow region of -2.9 Sv in July–August which later relaxed to an eastward transport of 0.8–1.0 Sv. This reversal of a long-slope current is confirmed by independent observations from NABOS moorings.

b. Formation of a prominent SSH trough extending from the eastern Laptev Sea to the Bering Strait. A similar and even stronger structure was obtained in the PIOMAS solution and is indirectly evidenced by two NABOS moorings located on the continental slope of the Laptev Sea.

c. The aforementioned SSH depression near the Chukchi Sea tends to increase the large-scale sea level difference between the Bering Sea and the AO. This contributes to the 25% increase in the Bering Strait transport at that time and agrees with the regional force balance suggesting an increased role of the pressure head between the Bering Sea and AO during 2007–2011 [20].

d. A significant total FWC of $\sim$20,700 km^3 in the BG during 2008. The FW accumulation agrees with estimates from in situ hydrographic observations [46]. Analysis of the FW transports across model boundaries around the BG indicates that FW accumulation in 2008 was mainly caused by the anomalous inflow through the eastern section. The DA model estimate of $\sim$0.8 Sv qualitatively agrees with other works [58, 59] that suggest FW sources may include from near Greenland.

Acknowledgements

J. N. Stroh thanks the University of Nevada, Reno DeLaMare Library, for document preparation resources. G. Panteleev and M. Yaremchuk were supported by the Office of Naval Research (ONR) project "Arctic data assimilation," Program Element 0602435N. O. Francis was supported by the Coastal Hydraulics Engineering Resilience (CHER) Lab, Department of Civil and Environmental Engineering, and National Sea Grant College Program at the University of Hawaii at Manoa. M. Yaremchuk was also supported by the ONR Program Element 0603207N under work on the Navy Earth System Prediction Capability.

Author details

J.N. Stroh[1*], S. Kirillov[2], G. Panteleev[3], O. Francis[4], M. Yaremchuk[3], E. Bloshkina[5] and N. Lebedev[5]

1 Independent Scholar, Fairbanks, AK, USA

2 University of Manitoba, Winnipeg, Canada

3 Naval Research Laboratory, Stennis Space Center, MS, USA

4 University of Hawaii at Manoa, Honolulu, HI, USA

5 Arctic and Antarctic Research Institute, St. Petersburg, Russia

*Address all correspondence to: j.nate.stroh@gmail.com

IntechOpen

References

[1] Rudels B, Jones EP, Anderson LG, Kattner G. On the intermediate depth waters of the Arctic Ocean. In: The polar oceans and their role in shaping the global environment. Washington, D.C.: American Geophyscial Union; 1994. pp. 33-46

[2] Zhang X, Zhang J. Heat and freshwater budgets and pathways in the Arctic Mediterranean in a coupled ocean/sea-ice model. Journal of Oceanography. 2001;57(2):207-234

[3] Shiklomanov IA, Shiklomanov AI, Lammers RB, Peterson BJ, Vorosmarty CJ. The dynamics of river water inflow to the Arctic Ocean. In: The Freshwater Budget of the Arctic Ocean. Dordrecht: Springer; 2000. pp. 281-296

[4] Serreze MC, Barrett AP, Slater AG, Woodgate RA, Aagaard K, Lammers RB, et al. The large-scale freshwater cycle of the Arctic. Journal of Geophysical Research: Oceans. 2006;111:1-19. C11010

[5] Curry B, Lee CM, Petrie B. Volume, freshwater, and heat fluxes through Davis Strait, 2004–05. Journal of Physical Oceanography. 2011;41(3):429-436

[6] Weingartner TJ, Cavalieri DJ, Aagaard K, Sasaki Y. Circulation, dense water formation, and outflow on the northeast Chukchi shelf. Journal of Geophysical Research: Oceans. 1998;103(C4):7647-7661

[7] Dmitrenko IA, Rudels B, Kirillov SA, Aksenov YO, Lien VS, Ivanov VV, et al. Atlantic water flow into the Arctic Ocean through the St. Anna trough in the northern Kara Sea. Journal of Geophysical Research: Oceans. 2015;120(7):5158-5178

[8] Rudels B, Friedrich HJ, Quadfasel D. The Arctic circumpolar boundary current. Deep Sea Research Part II:

Topical Studies in Oceanography. 1999;46(6–7):1023-1062

[9] Proshutinsky AY, Johnson MA. Two circulation regimes of the wind-driven Arctic Ocean. Journal of Geophysical Research: Oceans. 1997;102(C6):12493-12514

[10] Woodgate R. Arctic Ocean circulation: Going around at the top of the world. Nature Education Knowledge. 2013;4(8):8

[11] Dukhovskoy D, Johnson M, Proshutinsky A. Arctic decadal variability from an idealized atmosphere-ice-ocean model: 2. Simulation of decadal oscillations. Journal of Geophysical Research: Oceans. 2006;111:1-17. C06029

[12] Polyakov IV, Johnson MA. Arctic decadal and interdecadal variability. Geophysical Research Letters. 2000;27(24):4097-4100

[13] Wang J, Zhang J, Watanabe E, Ikeda M, Mizobata K, Walsh JE, et al. Is the dipole anomaly a major driver to record lows in Arctic summer sea ice extent? Geophysical Research Letters. 2009;36(5):1-5. L05706

[14] Shiklomanov AI, Lammers RB. Record Russian river discharge in 2007 and the limits of analysis. Environmental Research Letters. 2009;4(4):1-9. 045015

[15] Polyakov IV, Beszczynska A, Carmack EC, Dmitrenko IA, Fahrbach E, Frolov IE, et al. One more step toward a warmer Arctic. Geophysical Research Letters. 2005;32(17):1-4. L17605

[16] McPhee MG, Proshutinsky A, Morison JH, Steele M, Alkire MB. Rapid change in freshwater content of the

Arctic Ocean. Geophysical Research Letters. 2009;**36**(10):1-6. L10602

[17] Polyakov IV, Alexeev VA, Ashik IM, Bacon S, Beszczynska-Möller A, Carmack EC, et al. Fate of early 2000s Arctic warm water pulse. Bulletin of the American Meteorological Society. 2011; **92**(5):561-566

[18] Morison J, Kwok R, Peralta-Ferriz C, Alkire M, Rigor I, Andersen R, et al. Changing arctic ocean freshwater pathways. Nature. 2012;**481**(7379):66

[19] Falck E, Kattner G, Budéus G. Disappearance of Pacific water in the northwestern Fram Strait. Geophysical Research Letters. 2005;**32**(14):1-4. L14619

[20] Woodgate RA, Weingartner TJ, Lindsay R. Observed increases in Bering Strait oceanic fluxes from the Pacific to the Arctic from 2001 to 2011 and their impacts on the Arctic Ocean water column. Geophysical Research Letters. 2012;**39**(24):1-6. L24603

[21] Carmack E, McLaughlin F, Yamamoto-Kawai M, Itoh M, Shimada K, Krishfield R, et al. Freshwater storage in the Northern Ocean and the special role of the Beaufort Gyre. In: Arctic–SubArctic Ocean Fluxes. Dordrecht: Springer; 2008. pp. 145-169

[22] Alkire MB, Falkner KK, Rigor I, Steele M, Morison J. The return of Pacific waters to the upper layers of the Central Arctic Ocean. Deep Sea Research Part I: Oceanographic Research Papers. 2007;**54**(9):1509-1529

[23] Zhong W, Zhao J. Deepening of the Atlantic water core in the Canada Basin in 2003–11. Journal of Physical Oceanography. 2014;**44**(9):2353-2369

[24] Steele M, Boyd T. Retreat of the cold halocline layer in the Arctic Ocean.

Journal of Geophysical Research: Oceans. 1998;**103**(C5):10419-10435

[25] Emery WJ. Water types and water masses. In: Encyclopedia of Ocean Sciences. Vol. 6. San Diego: Academic Press; 2001. pp. 3179-3187

[26] Stroh JN, Panteleev G, Kirillov S, Makhotin M, Shakhova N. Sea-surface temperature and salinity product comparison against external in situ data in the Arctic Ocean. Journal of Geophysical Research: Oceans. 2015; **120**(11):7223-7236

[27] Steele M, Morley R, Ermold W. PHC: A global ocean hydrography with a high-quality Arctic Ocean. Journal of Climate. 2001;**14**(9):2079-2087

[28] Environmental Working Group. In: Timokhov L, Tanis F, editors. Environmental Working Group Joint U.S.-Russian Atlas of the Arctic Ocean, Version 1. Boulder, Colorado USA: NSIDC: National Snow and Ice Data Center; 1997. DOI: 10.7265/N5H12ZX4. Available from: http://nsidc.org/data/g01961

[29] Francis OP, Yaremchuk M, Panteleev GG, Zhang J, Kulakov M. Anomalous circulation in the Pacific sector of the Arctic Ocean in July–December 2008. Ocean Modelling. 2017; **117**:12-27

[30] Toole JM, Krishfield RA, Timmermans ML, Proshutinsky A. The ice-tethered profiler: Argo of the Arctic. Oceanography. 2011;**24**(3):126-135

[31] Krishfield R, Toole J, Proshutinsky A, Timmermans ML. Automated ice-tethered profilers for seawater observations under pack ice in all seasons. Journal of Atmospheric and Oceanic Technology. 2008;**25**(11): 2091-2105

[32] Troupin C, Barth A, Sirjacobs D, Ouberdous M, Brankart JM, Brasseur P,

et al. Generation of analysis and consistent error fields using the data interpolating variational analysis (DIVA). Ocean Modelling. 2012;**52**: 90-101

[33] Jakobsson M, Mayer L, Coakley B, Dowdeswell JA, Forbes S, Fridman B, et al. The international bathymetric chart of the Arctic Ocean (IBCAO) version 3.0. Geophysical Research Letters. 2012;**39**(12):1-6. L12609

[34] Ekwurzel B, Schlosser P, Mortlock RA, Fairbanks RG, Swift JH. River runoff, sea ice meltwater, and Pacific water distribution and mean residence times in the Arctic Ocean. Journal of Geophysical Research: Oceans. 2001; **106**(C5):9075-9092

[35] Steele M, Morison J, Ermold W, Rigor I, Ortmeyer M, Shimada K. Circulation of summer Pacific halocline water in the Arctic Ocean. Journal of Geophysical Research: Oceans. 2004; **109**:1-18. C02027

[36] Shimada K, Itoh M, Nishino S, McLaughlin F, Carmack E, Proshutinsky A. Halocline structure in the Canada Basin of the Arctic Ocean. Geophysical Research Letters. 2005;**32**(3):1-5. L03605

[37] Shimada K, Carmack EC, Hatakeyama K, Takizawa T. Varieties of shallow temperature maximum waters in the western Canadian Basin of the Arctic Ocean. Geophysical Research Letters. 2001;**28**(18):3441-3444

[38] Bourgain P, Gascard JC. The Atlantic and summer Pacific waters variability in the Arctic Ocean from 1997 to 2008. Geophysical Research Letters. 2012;**39**(5):1-6. L05603

[39] McLaughlin FA, Carmack EC, Williams WJ, Zimmermann S, Shimada K, Itoh M. Joint effects of boundary currents and thermohaline intrusions on

the warming of Atlantic water in the Canada Basin, 1993–2007. Journal of Geophysical Research: Oceans. 2009; **114**:1-20. C00A12

[40] Polyakov IV, Timokhov LA, Alexeev VA, Bacon S, Dmitrenko IA, Fortier L, et al. Arctic Ocean warming contributes to reduced polar ice cap. Journal of Physical Oceanography. 2010; **40**(12):2743-2756

[41] Morison J, Steele M, Kikuchi T, Falkner K, Smethie W. Relaxation of Central Arctic Ocean hydrography to pre-1990s climatology. Geophysical Research Letters. 2006;**33**(17):1-5. L17604

[42] Rabe B, Karcher M, Schauer U, Toole JM, Krishfield RA, Pisarev S, et al. An assessment of Arctic Ocean freshwater content changes from the 1990s to the 2006–2008 period. Deep Sea Research Part I: Oceanographic Research Papers. 2011;**58**(2):173-185

[43] Ge S, Yang D, Kane DL. Yukon River basin long-term (1977–2006) hydrologic and climatic analysis. Hydrological Processes. 2013;**27**(17): 2475-2484

[44] Zhang T, Frauenfeld OW, Serreze MC, Etringer A, Oelke C, McCreight J, et al. Spatial and temporal variability in active layer thickness over the Russian Arctic drainage basin. Journal of Geophysical Research: Atmospheres. 2005;**110**:1-14. D16101

[45] Shiklomanov AI, Lammers RB, Rawlins MA, Smith LC, Pavelsky TM. Temporal and spatial variations in maximum river discharge from a new Russian data set. Journal of Geophysical Research: Biogeosciences. 2007;**112**:1-14. G04S53

[46] Proshutinsky A, Krishfield R, Timmermans ML, Toole J, Carmack E, McLaughlin F, et al. Beaufort Gyre

freshwater reservoir: State and variability from observations. Journal of Geophysical Research: Oceans. 2009; **114**:1-25. C00A10

[47] De Steur L, Steele M, Hansen E, Morison J, Polyakov I, Olsen SM, et al. Hydrographic changes in the Lincoln Sea in the Arctic Ocean with focus on an upper ocean freshwater anomaly between 2007 and 2010. Journal of Geophysical Research: Oceans. 2013; **118**(9):4699-4715

[48] Curry B, Lee CM, Petrie B, Moritz RE, Kwok R. Multiyear volume, liquid freshwater, and sea ice transports through Davis Strait, 2004–10. Journal of Physical Oceanography. 2014;**44**(4): 1244-1266

[49] Beszczynska-Möller A, Woodgate RA, Lee C, Melling H, Karcher M. A synthesis of exchanges through the main oceanic gateways to the Arctic Ocean. Oceanography. 2011;**24**(3):82-99

[50] McPhee MG. Intensification of geostrophic currents in the Canada Basin, Arctic Ocean. Journal of Climate. 2013;**26**(10):3130-3138

[51] Armitage TW, Bacon S, Kwok R. Arctic Sea level and surface circulation response to the Arctic oscillation. Geophysical Research Letters. 2018;**45**: 6576-6584

[52] Panteleev G, Yaremchuk M, Stabeno PJ, Luchin V, Nechaev DA, Kikuchi T. Dynamic topography of the Bering Sea. Journal of Geophysical Research: Oceans. 2011;**116**:1-11. C05017

[53] Luchin V, Panteleev G. Thermal regimes in the Chukchi Sea from 1941 to 2008. Deep Sea Research Part II: Topical Studies in Oceanography. 2014;**109**: 14-26

[54] Panteleev G, Nechaev DA, Proshutinsky A, Woodgate R, Zhang J. Reconstruction and analysis of the

Chukchi Sea circulation in 1990–1991. Journal of Geophysical Research: Oceans. 2010;**115**:1-22. C08023

[55] Timmermans ML, Proshutinsky A, Krishfield RA, Perovich DK, Richter-Menge JA, Stanton TP, et al. Surface freshening in the Arctic Ocean's Eurasian Basin: An apparent consequence of recent change in the wind-driven circulation. Journal of Geophysical Research: Oceans. 2011;**116**: 1-17. C00D03

[56] Semiletov I, Dudarev O, Luchin V, Charkin A, Shin KH, Tanaka N. The east Siberian Sea as a transition zone between Pacific-derived waters and Arctic shelf waters. Geophysical Research Letters. 2005;**32**(10):1-5. L10614

[57] Maslanik J, Drobot S, Fowler C, Emery W, Barry R. On the Arctic climate paradox and the continuing role of atmospheric circulation in affecting sea ice conditions. Geophysical Research Letters. 2007;**34**(3):1-4. L03711

[58] Lique C, Garric G, Treguier AM, Barnier B, Ferry N, Testut CE, et al. Evolution of the Arctic Ocean salinity, 2007–08: Contrast between the Canadian and the Eurasian basins. Journal of Climate. 2011;**24**(6): 1705-1717

[59] Proshutinsky A, Dukhovskoy D, Timmermans ML, Krishfield R, Bamber JL. Arctic circulation regimes. Philosophical Transactions of the Royal Society A. 2015;**373**(2052):1-18. 20140160

Chapter 3

Coastal Erosion Due to Decreased Ice Coverage, Associated Increased Wave Action, and Permafrost Melting

Dimitrios Kostopoulos, Efrat Yitzhak and Ove T. Gudmestad

Abstract

It is broadly recognized that the Arctic area has become highly popular for hosting new activities and new infrastructure. This is due to the combination of the need of exploring new areas to satisfy the ever increased energy demand and also the impact of climate change that has created paths for increased trading and maritime activities. Presently, the Arctic environment poses new challenges and unknown hazards, which are considered unpredictable due to the uncertainties of the emerging phenomena. In this chapter, the effects caused by the higher temperatures in the Arctic region on the increased height of waves and storm surges and the extended erosion of the Arctic coastline are examined and presented. This unpredictability is partly due to the dynamic behavior of the Arctic environment and the annual fluctuations of the permanent ice of the Arctic Ocean. Reduced ice coverage, especially during the fall period, creates longer available sea distances for waves to be developed. As extreme case scenario, the associated consequences for the design wave height on a totally ice-free sea are studied. A comparison between the heights of the waves which are generated by the longest possible fetches and those estimated from today's ice limit situation is made based on coastal engineering methods. Further to this, more open sea areas also allow for increased storm surge heights. In the chapter, it is also shown how the decreased ice coverage has an influence on the coastal erosion phenomenon, which is not only enhanced due to the evolving wave dynamics but also thermodynamics and sediment dynamics. The presented results show significant changes of the characteristic wave heights and strong increase of the pace of the coastal erosion. Based on these observations, the authors of this chapter want to stress the challenges that such future conditions in the Arctic area will pose to any Arctic operations, nearby infrastructures and human activities in the area.

Keywords: Arctic Ocean, free ice sea, extreme wave heights, permafrost melting, shore erosion

1. Introduction

The Arctic physical environment is characterized by various dynamic phenomena, sudden ones, like polar lows and unexpectedly strong storms, or time developing and periodical, like gradual coastal erosion of the shoreline. In order to operate safely in this environment, one needs to be undoubtedly supported by daily weather forecasting and monitoring. However, accurate means of doing so and good prognostics are challenged by the lack of historical and scientific data as well as a limited number of stations for data collection, which make the Arctic Ocean a hazardous environment with challenging marine and weather conditions.

Recent events testify the aforementioned hazardousness. For example, on July 24, 2010, in the Varandey area in northern Russia, the oil treatment and storage terminal located kilometers inland was flooded and the airport runway closed, due to the fact that the coast was severely damaged by excessive flooding. This flooding event was the outcome of combined storm waves, surges, and tides. Other northern production sites, such as the Northstar artificial oil and gas production island in the Beaufort Sea, have also been damaged by significantly high waves. In that case during the design phase, the facilities, which are located 19 km northwest of Prudhoe Bay, Alaska and 10 km north of the Alaskan coast at a water depth of 10 m, were designed using historical data and assumptions of fetch length and wave height occurrence which did not correspond to events that happened some years after production startup.

In this chapter, we are analyzing some of these challenges and phenomena, taking into consideration the significant changes that have occurred in the Arctic area during the last decades. For instance, throughout the years, the average monthly Arctic sea ice extent has dropped dramatically from 12.5 million km^2 in 1980s to about 10.8 million km^2 in 2016, showing a declining trend of 4.1% per decade (see **Figure 1**) [1]. This means that at coastlines and areas that before used to be covered by snow permanently, people now observe waves up to 4 m in height. Due to the retraction of the ice cover, new paths for trading and transportation are seasonally opened, like the North Sea Route (the Northeastern Passage), which is now used as a transport path with ships for liquefied natural gas (LNG) from the Sabetta LNG facilities on Yamal to the Chinese market. During the summer period and early autumn, when the passage is almost ice free, operators can travel from Europe to Asia using this path to the north of Russia with the service of icebreakers.

The wave forces that are generated due to the ice-free surface enhance the ice shrinkage and reduce the ice thickness, helping ice edges to detach more easily from the main ice core. Another observation is the increase of the temperature and seasonal record peaks that might be also a consequence of the annual shrinkage of the permanent ice extent which works as natural mirror and shield against the heat. The increase of the temperature does consequently lead to increased ice melting creating a loop of domino effects.

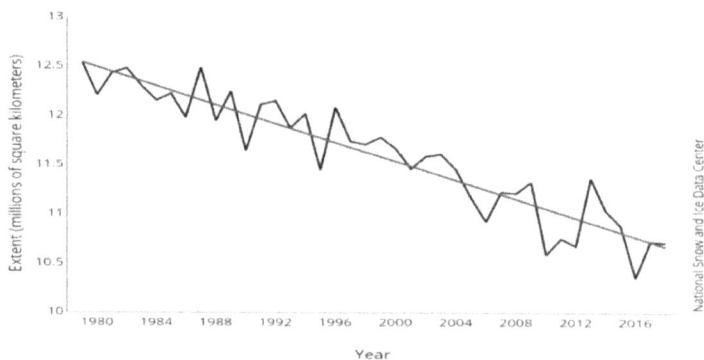

Figure 1.
Monthly June Arctic sea ice extent for 1979–2018 shows a decline of 4.1% per decade [1].

All the aforementioned changes testify to a likely future situation where ice surface will shrink further and possibly occasionally disappear. Shipping and operations in the area will face a different environment of what they are designed for today. Some new challenges might occur. As long as waves are considered, it is probable that assets will face a more hazardous environment with higher waves generated in an open ocean. Some business people might claim that conditions will be more favorable for operations in the Arctic if temperature increases, since this will alleviate winterization issues. However, no one can predict with certainty what the environmental conditions will be and how the aforementioned changes will influence the phenomena by reducing or increasing some risks.

In relation to the definition of the risk of an activity, A, is the multi-dimensional combination of its probability P, its consequences, C, and the related uncertainties, U, of what the outcome will be (A, P, C, and U). The uncertainty of the activity is well linked to the knowledge that one has about the activity. Therefore, since in the Arctic area there is lack of knowledge due to scarce historical data or measurements of previous hazardous events for a sufficient long period, risks can be considered inherently high in the Arctic area where safety for the assets or humans may not be guaranteed.

Thus, there is a need to understand better the challenges that might occur in the future by assessing some potential future scenarios. One such scenario is an open Arctic Ocean where there is no ice. In this chapter, this scenario is related to the potential increase of the wave height in specific areas. One specific method for predicting maximum wave heights is used, here, covering the subject briefly and giving food for further research and analysis.

Winds blowing over the sea generate ocean surface waves (wind-sea and swell) which are related to the distance (length of fetch) and the duration of wind. As both wind-sea and swells depend on the open water sea fetch, further reductions in seasonal ice cover will result in larger waves [2].

Such larger waves can have multiple consequences to the coasts around but also to the marine operations in the area. Wave activity when reaching the shallow areas along the coast leads to currents and water circulation that can cause excessive erosion and enhanced sediment transportation. Also, present navigation experience

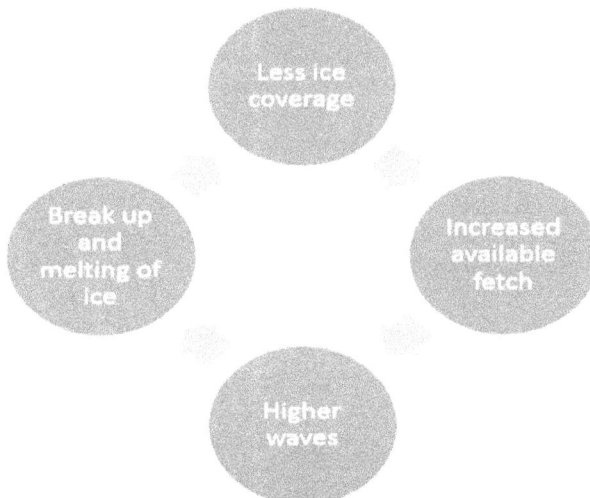

Figure 2.
Feedback loop of the wave ice interaction.

can be challenged due to higher waves generated by rapid storms and changing seafloor conditions. In the future Arctic Ocean, wave conditions like those will be changing the known environment for nature and humans.

Moreover, the existence of ice on the sea surface makes the phenomenon of wave and ice interaction complex. Ice masses suppress waves, diminishing them, but also waves alter and influence the thickness and the growth of the ice. Waves start penetrates more and more into the weakened sea-ice reaching the marginal ice zone, the part of the ice cover that interacts with the open ice-free ocean. This loop produces a positive feedback that could accelerate the loss of ice especially during summer and early fall [3] (**Figure 2**).

2. Methodology

As mentioned before, the aim of this chapter is not to execute an excessive assessment of the ice melting impact on the design wave heights, but rather stress and highlight the challenge that might occur in a worst-case scenario when there is no ice in the Arctic. In this way, one can have a better understanding of the magnitude of change that could be expected. These effects are also present, possibly in a less extend and locally, when there is a partial reduction of the ice surface and not total disappearance.

The methodology chosen is based on the assumption that the Arctic Ocean in a free-ice period can be considered as a gigantic ocean, surrounded by the northern coasts of the neighboring countries. When one aims to estimate the characteristic wave height, two factors are the main contributors that need to be taken into consideration. One is the fetch length, the length of water over which a given wind can blow, and this is also the main factor that creates storm surge which also leads to coastal erosion and flooding. The other factor is the wind characteristics, such as duration and velocity. Thus, focus initially was given on areas that have the longest potential fetch distance, assuming that the wind conditions (duration and speed) are similar from all directions.

Taking all the aforementioned factors into consideration and also the available data and their quality, the northern part of Svalbard Island is chosen to be examined. As per now, this part of the Arctic region is covered by ice during most of the year, but in a free ice future scenario, long fetches are revealed that can potentially generate high waves. Moreover, statistical studies have showed that the percentage of north easterly winds occurring annually at the examined area is significant which testifies the relevance of the choice of location and direction for the study. Five different meteorological stations at the north and northeast area of Svalbard are selected to acquire the desired wind data. These data are analyzed to extract information regarding the most extreme wind incidents and storms from 2000 to 2014. The collected data refer to the early fall period of the year (September and October), as this is the period when the Arctic is expected to have the lowest ice percentage.

The stations are as follows [4]:

- KARL XII (99935): Latitude: 80.653, Longitude: 25.008

- KONGSØYA (99740): Latitude: 78.9277, Longitude: 28.892

- VERLEGENHUKEN (99927): Latitude: 80.059, Longitude: 16.25

- KVITØYA (99938): Latitude: 80.07, Longitude: 31.5

- HOPEN (99720): Latitude: 76.5097, Longitude: 25.0133

Since the main focus of this study is to examine the difference in wave height estimates due to the shrinkage of ice coverage, the directions that are examined are those that showed the most dramatic change in length. Thus, for the examined area, the directions that are chosen are all between 320 and 55°. As shown in **Figure 3**, in the case of ice coverage disappearance, the increase of the fetch length in some directions is up to three times longer than the one today (from presently 500 to 2000 km to the northern coast of Russia and from 200–250 to 1500–3000 km to the coasts of Canada and Alaska).

The calculation of the maximum characteristic wave height is made by using the Jonswap method. This method is chosen as it is judged appropriate for open sea waves and considers the influence of the fetch length. According to this method, a wind generated wave can be either fetch limited (limited by the available distance over which it has been generated) or duration limited (limited by the period of time that the wind is blowing).

In order to find the largest wave height that can occur due to the wind phenomena in the region, all the characteristic significant waves, H_s, for the directions of interest were assessed. In Svalbard, the examined directions were NNW, N, and NNE (North-Northwest 340°, North 0°, and North-Northeast 32.5°). Those directions were chosen because we wanted to cover as much as we could of the examined area for three different directions. For that reason, straight lines were drawn for each 2.5° with the use of maps from Google Earth (see **Figure 3**) until the opposite coasts were reached.

For each and every one of the three wind directions that were examined (NNW, N, and NNE), wind was assumed from an angle of 45°, that is, from −22.5 to +22.5° for each direction (usually the spreading used is 90°, but here, because of the limited examined area, we had to choose a smaller and more narrow area, the half). The fetch length, F, was drawn for every α_i = 2.5° angle around each direction, and they are calculated by the following quation [5]:

$$F = \frac{\sum_{i=-N}^{N} F_i \cos^2 a_i}{\sum_{i=-N}^{N} F_i \cos^2 a_i} \tag{1}$$

where N is the number of each fetch line drawn between −22.5 and + 22.5° for each of the three directions (NNW, N, and NNE).

The examined directions were chosen based on the morphology of the area; the islands and coasts of Greenland at the northwest part, for example, do not allow the development of long enough fetches to be considered.

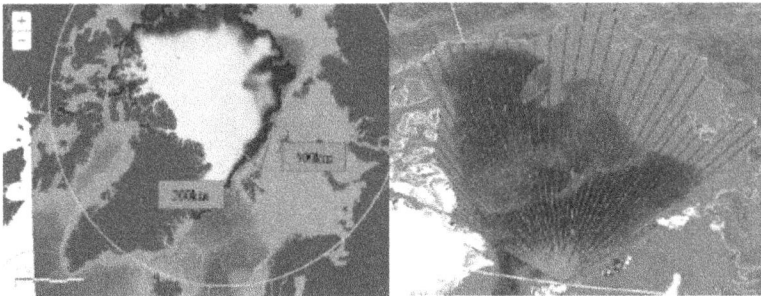

Figure 3.
Present maximum and minimum available fetch lengths north of Svalbard during early autumn periods (left). Future fetch lengths available in a free ice Arctic Ocean, between 320 and 55° in steps of 2.5°.

As long as the wind duration is considered, it is important to mention, here, that it would have been reasonable to have data with annual percentage of occurrence for each wind velocity in each direction. However, such data were not available; therefore, all the calculations are performed for specific events collected by the five stations over the past 15 years. So, for every fetch direction, one average wind speed is calculated. This means that every storm observed in the data is related to the three main directions (NNW, N, and NNE), and a mean wind velocity is calculated which is used to describe the wind at 10 m altitude.

Steps of the Jonswap method [5]:

Step 1. Calculation of the frictional wind velocity

$$u_* = W\sqrt{0.001(1.1 + 0.035W)} \qquad (2)$$

where W = mean velocity at 10 m height.

Step 2. Calculation of the equivalent fetch, F_{eq}, depending on the duration of the wind:

$$\frac{g\,F_{eq}}{u_*} = 0.00523\left(\frac{g\,t_d}{u_*^2}\right)^{1.5} \qquad (3)$$

where g = gravity acceleration, 9.81 m/s^2; t_d = the duration of the wind blowing; F_{eq} = the equivalent fetch length.

Step 3. Checking whether the wave is duration or fetch limited:

- If $F_{eq} > F$, then the wave is fetch limited, and the fetch, F, of the specific direction (Eq. (1)), needs to be used for the calculation of the characteristic height, H_s.

- If $F_{eq} < F$, then the wave is duration limited, and the F_{eq} should be used for the height calculation.

Step 4. Calculation of the characteristic wave height

$$\frac{g\,H_s}{u_*^2} = 0.0413\left(\frac{gF}{u_*^2}\right)^{0.5} \qquad (4)$$

$$\frac{g\,H_s}{u_*^2} = 0.0413\left(\frac{g\,F_{eq}}{u_*^2}\right)^{0.5} \qquad (5)$$

H_s = characteristic wave height.

Step 5. Calculation of the characteristic period of the wave

$$\frac{g\,T_s}{u_*} = 0.71345\left(\frac{gF}{u_*^2}\right)^{0.33} \qquad (6)$$

T_s = characteristic wave period.

3. Results

Based on the previous methodology, the results of the calculation are as shown below. **Table 1** is showing the maximum possible fetch distances, as calculated using

Final fetches for each direction	
Direction	Fetch F (km)
North-Northeast	2661
North	3130
North-Northwest	2494

Table 1.
Final fetch lengths for each examined direction in a future ice-free Arctic.

Significant wave height and characteristic period							
Wind direction	W (m/s)	t_d (hours)	u*	F_{eq} (km)	Comment	H_s (m)	T_s (s)
NNW	16.60	90	0.68	630.51	Duration limited	7.13	11.12
N	11.38	102	0.44	2418.72	Duration limited	9.03	14.94
NNE	14.55	36	0.58	2307.72	Duration limited	11.70	16.20

Table 2.
Final wave significant heights and characteristic periods in a future ice free Arctic.

Significant wave height and characteristic period							
Wind direction	W (m/s)	t_d (hours)	Fetch (km)	F_{eq} (km)	Comment	H_s (m)	T_s (s)
NNW	16.60	90	160	630.51	Fetch limited	3.59	11.12
N	11.38	102	150	2418.72	Fetch limited	2.25	14.94
NNE	14.55	36	350	2307.72	Fetch limited	4.52	16.20

Table 3.
Final wave significant heights and characteristic periods in today's conditions.

Eq. (1). This fetch is as expected in the case of an ice-free Arctic Ocean. Based on the aforementioned available fetch, the significant height of the waves together with the characteristic wave period was calculated and shown in **Table 2**.

Since the aim of the chapter is to compare the waves of the future ice-free scenario with those of today, wave characteristic heights based on current conditions are calculated and shown in **Table 3**.

All in all, the results show a significant increase of the height in the case of an ice-free Arctic Ocean. Actually, since the waves were duration limited, it is possible that such waves can be generated even with some permanent ice coverage. In detail, in the NNW direction, the wave height was almost doubled, from 3.59 to 7.13 m. In the North direction, the most significant change is observed with more than three times magnification of the characteristic height, from 2.25 to 9.03 m. Last, in the NNE direction, the prediction shows an increase from 4.52 to 11.70 m.

4. Discussion

The examined scenario of ice retraction should not be considered as topical only in the Svalbard area. Many measurements and experimental campaigns have been made in the eastern part of the Arctic Ocean, close to Beaufort Sea, where the sea ice cover has retreated significantly. Due to this dramatic retreat, especially in September 2012, 5 m height waves were observed in the middle of the basin. These were extremely large waves compared to what has been observed previously, testifying the assumption and the prediction of wave height enhancement due to ice surface shrinkage [2].

Apart from experimental campaigns and measurements, other studies using prognostic models have shown significant changes in estimated wave heights. These changes are undoubtedly linked to the increase of the fetch length created by the free-ice sea area. What is worth mentioning here is that the results showed also a rise in surface winds in the Arctic area, mainly in Kara, Laptev, and East Siberian Seas. On the contrary, at the western part of the Arctic region, in the Barents Sea, a drop of the winds and consequently the wave heights were observed [6].

Moreover, research supports the assumption that in areas where the ice coverage is shrinking, the wave phenomena will change. Results have shown a growth in wind speeds and an increase of the frequency of occurrence of waves of 2 m height. On the other hand, the same studies have shown that the change in extreme wave heights is marginal. The areas where the change is more significant are those of the northern parts of Barents Sea, Kara, and Chukchi Seas, whereas, in areas where the sea is already ice free during September and October, like the North Atlantic and the main part of the Barents Sea, extreme waves would be less frequently witnessed and great changes in extreme wave heights could not be expected [7]. In conclusion, the eastern Arctic regions and areas close to the north Canadian coasts will be influenced most by the absence of the ice [6].

It should be noted that the discussion above relates to wave heights only. To estimate the sea level during a storm surge in the case of wind in direction toward shore has been outside our scope. A storm surge that encounters a shallow shore could climb up the coast easily, causing floods and increased erosion, while a storm surge that approaches a steep shore is more likely to break early, thus, cliff or steep shore might be sufficient obstacle to prevent a storm surge from piling-up and reaching far inland. The combined storm surge and waves will cause flooding and damages far inland. An unprecedented amount of erosion could occur due to the effect of flooding and wave action, in particular, as higher temperatures cause the increased melting of permafrost along the shores, these effects will be discussed below.

5. The melting of the permafrost

Warmer climate and rising temperatures affect the Arctic in many aspects. Thawing permafrost is of the phenomenon that is detected in the Arctic. Measurements over long periods of time show that the permafrost temperature has rose by up to 2°C, and shallow permafrost layers in some areas have thawed completely. Consequently, the permafrost extent has shrunk by 30–80 km in Russia and up to 130 km in Canada [8]. In addition, a decrease in the snow cover creates a feedback mechanism of increasing temperatures. These phenomena lead to unstable grounds and emissions of greenhouse gases and toxicants that had been encapsulated in the frozen ground, and the permafrost is thawing in areas which were permanently frozen until recently [9]. The Arctic shores easily erode when hit by storm surges and strong waves. As a result of melted materials being washed away, the shore becomes even more susceptible to erosion [10].

The permafrost's thermal properties, conductivity, heat capacity, thermal diffusivity, latent heat, and thermal expansion, are among the key variables in determining permafrost melting and erosion rate. The thawing rate of a given soil depends on the soil composition (soil particles, ice, water, and air content of the soil) and the conditions of the physical environment. By knowing the content of frozen water in the ground and combining it with the assumptions that:

- Ice melts at temperatures above 0°C.

- The soil's thawing temperature is 0°C.

- All melted ground will wash away and erode by the impact of storm surge and waves.

It is possible to roughly estimate the amount of ground that will erode. When trying to assess coastal erosion, many uncertainties and parameters must be taken into consideration [10]. It was also necessary to make some simplifications and assumptions in order to get a model which can predict soil temperature.

An important parameter is the Degree days, the product of temperature and number of days. The degree days for an average temperature of 3°C over a period of 7 days is, for example, $3 \cdot 7 = 21°C \cdot d$. The thawing index (I_{st}) is the number of degree days where the temperature is above the melting temperature (for water, 0°C). To calculate I_{st}, the degree days of each month were calculated: a monthly average temperature was calculated and multiplied by the number of days in each month. Under the assumption that the soil melting temperature is 0°C, I_{st} of the soil is the summation of degree days above 0°C for a one-year period.

The thermal models for coastal erosion that we used are described in [11]. For thawing depth estimation, first an evaluation of the permafrost soil consistency (soil profile and water content) was made, and then we were using the Stefan's equation (see [9–11]) to estimate the thawing depth in a partly frozen soil.

6. Permafrost erosion models

Several assumptions were made for estimating the amount of eroded soil during the one year period:

- Erosion occurs between May and September (the erosion process is negligible between October and April, due to sea ice and frozen soil).

- A big storm surge hits the shores at the end of each season (Spring, Summer, and Autumn) and erodes all melted soil.

- A season would count as a 50-day period.

As the average erosion rate in Varandey area was 2.7 m/year between 2005 and 2007 [11], based on these assumptions, the amount of eroded soil could be estimated.

The assumption that all melted material is being removed by a storm surge at the end of each "season" means that a new frozen soil layer is now exposed to heat and melting processes. A melted soil layer that stays intact could create an insulation layer that prevents heat penetration and decreases the melting processes, so the overall melted and eroded soil amount would be much smaller. For example, a single storm surge that hits the shore at the end of fall would hardly influence the erosion rate.

An erosion rate sensitivity analysis was made to assess and better understand the effect of the number of storms in a year on the total erosion rate. Three different

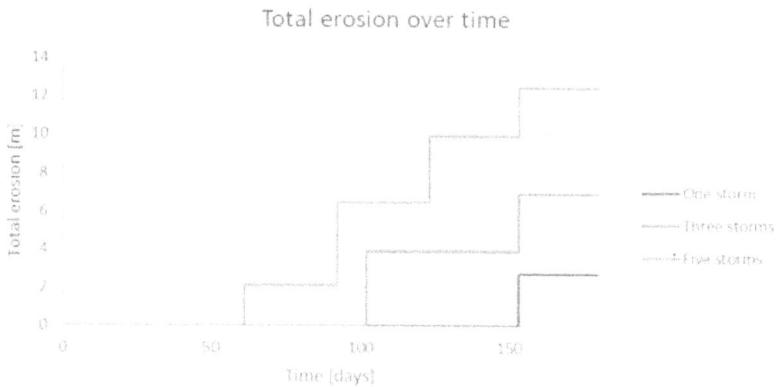

Figure 4.
The effect of number of storms on the total erosion as a function of time [10].

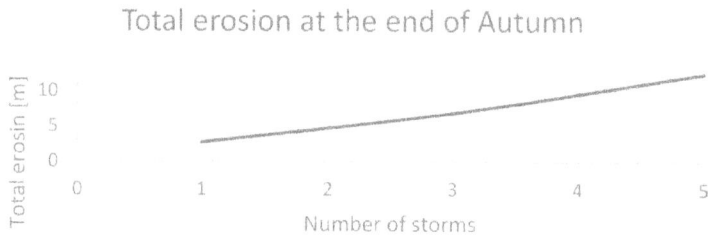

Figure 5.
Total erosion as a function of number of storm surges [10].

cases of storm surge were examined: the period between May and September was divided into sub-periods. It was then assumed that a storm surge hits the shore at the end of each sub-period and erodes all melted material.

As can be seen in **Figure 4**, the results correspond to the expectations—the erosion rate is increased as the number of storms rises. This result is further detailed in **Figure 5**, which shows the total erosion as number of storms/storm surges.

7. Conclusions

Undoubtedly, people and operations are facing extreme challenges in the Arctic Ocean. From polar lows and sudden storms to icing and iceberg drifts. However, more and more often, people are coming across extreme waves and permafrost erosion to an extent that has never been witnessed before. One of the reasons of this change is believed to be the melting of the ice and the alteration of the physical environment in the Arctic area. Wind blows over larger areas of the sea surface which consequently leads to more extreme wave phenomena and coastal erosion. Additionally, the increase of the annual average temperature and the prolongment of the warm periods influence the aforementioned phenomena which consequently lead to an increasing coastal erosion.

In this chapter, it is shown how such an ice shrinkage can influence the development of the waves by increasing the fetch length that will generate the examined waves. Additional research supports the aforementioned assumption, since measurements have testified that increase of the wave heights in areas where the waves were relatively mild. Of course, the outcomes from such research activities vary and further studies are needed to get a better view of the situation. What one can be certain of, though, is that permanent ice surface shrinkage will create a different wave and wind environment in the Arctic area. It is hard to say which of these factors will be the dominant and influence more significantly the situation.

The consequences of higher storm surge levels and higher waves and increased wave forces can be unpredictable. For instance, for already existing oil and gas platforms, which were designed according to historical data, unknown wave phenomena can, irreversibly, threaten human lives and assets. Therefore, in the case of higher waves, operators need to execute a reassessment of the air gap of the platform to avoid deck slamming; likewise, the strength of structures and safety factors should be reconsidered by including the uncertainties generated by the physical environment.

Due to increased wave action and melting of permafrost, Arctic coastlines and coastal infrastructure would see an increased stress from enhanced erosion and sediment circulation when sediment transportation along the coast alters.

In practice, newly opened Arctic seas will boost and encourage trading and navigation in the region since they will provide new paths with significant economic benefits. Shipping and offshore activities in areas which today we struggle to develop would be possible, but uncertainties related to storms and associated waves will remain, unless further studies are not made. Hazards that occur in open oceans might occur in the Arctic as well. For example, such hazards could be tsunamis, generated by earthquakes and motions of the seabed.

These are some examples of threats that so far were sleeping in the sea under the permanent ice coverage. Now, with its excessive melting, all these threats start coming on to the surface, putting in danger coastlines, people, and operations in the region.

Acknowledgments

The first author would like to thank Dr. Athanassios A. Dimas, Professor in the Department of Civil Engineering at the University of Patras, for clarifying uncertainties during the research.

Author details

Dimitrios Kostopoulos*, Efrat Yitzhak and Ove T. Gudmestad
Department of Mechanical and Structural Engineering and Materials Science,
Faculty of Science and Technology, University of Stavanger, Stavanger, Norway

*Address all correspondence to: dimitriskostopoulos20@gmail.com

IntechOpen

References

[1] National Snow and Ice Data Center, NSIDC. Arctic Sea Ice News & Analysis. Available from: http://nsidc.org/arcticseaicenews/ [Accessed: July 4, 2018]

[2] Thomson J, Rogers WE. Swell and sea in the emerging Arctic Ocean. Geophysical Research Letters. 2014;**41**(9):3136-3140

[3] Dumont D, Squire V, Sandven S, Sagen H, Bertino L. Forecasting waves-in-ice for Arctic operators, Touch Briefings. 2011. Available from: https://www.nersc.no/sites/www.nersc.no/files/Dumont_Final.pdf

[4] Norwegian Meterological Institute, eKlima, Norwegian Meteorological Institute. Available from: http://sharki.oslo.dnmi.no/portal/page?_pageid=73,39035,73_39049&_dad=portal&_schema=PORTAL [Accessed: March 27, 2015]

[5] U.S. Army. Chapter II-2: Meteorology and wave climate. In: Coastal Engineering Manual (Engineer Manual 1110-2-11000). Washington, DC: U.S. Army Corps of Engineers; 2006

[6] Khon VC, Mokhov II, Pogarskiy FA, Babanin A, Dethloff K, Rinke A, et al. Wave heights in the 21st century Arctic Ocean. Geophysical Research Letters. 2014;**41**(8):2956-2961

[7] Orimolade AP, Haver S, Gudmestad OT. Estimation of extreme significant wave heights and the associated uncertainties: A case study using NORA10 hindcast data for the Barents Sea. Marine Structures. 2016;**49**:1-17

[8] AMAP. Arctic Climate Issues 2011: Changes in Arctic Snow, Water, Ice and Permafrost. SWIPA 2011 Overview Report. Oslo: Arctic Monitoring and Assessment Program (AMAP); 2012. Available from: https://www.amap.no/documents/doc/arctic-climate-issues-2011-changes-in-arctic-snow-water-ice-andpermafrost/129

[9] Andersland OB, Ladanyi B. An Introduction to Frozen Ground Engineering. New York: Chapman & Hall, Inc.; 1994

[10] Yitzak E. Permafrost shore erosion in a warmer climate [MSc thesis]. University of Stavanger; 2018

[11] Guégan E. Erosion of permafrost affected coasts: Rates, mechanisms and modelling [doctoral theses at NTNU, 2015:328]. Trondheim, Norway: NTNU—Norwegian University of Science and Technology; 2015

Chapter 4

Recent Geodynamics and Seismicity of the European Arctic

Evgeniy Rogozhin, Galina Antonovskaya, Irina Basakina and Natalia Kapustyan

Abstract

The recent seismicity and heat flow density are indicators of geodynamic processes. For the European Arctic, the information of recent earthquakes were generalized and compiled on the general seismic catalog for 1998–2017 based on the new seismic data from stations opened in the region from 2011 to 2016. The general database of the heat flow from different data sources is compiled to obtain its spatial distribution for the European Arctic region. The relationships of heat flow values and seismic activity are discussed for this region, and combined geological and geophysical lithosphere cross sections are made in the latitudinal and meridional directions. The most geodynamic active structures and zones of tectonic stress concentration are distinguished; there are Gakkel Ridge and the Svalbard anticline in the Barents Sea region. Weak seismic events were recorded in the Novaya Zemlya region that reflect manifestations of recent tectonic activity.

Keywords: sea geotectonics, seismic activity, heat flow, deep structure, earthquake catalog, seismic stations

1. Introduction

The European sector of the Arctic region is an element of a geodynamic system that includes the ancient Eurasian continent and intensely developing younger Arctic Ocean. Seismic, gravity and geothermal fields establish a structural-density inhomogenuity of the crystalline basement and sedimentary cover, show the nature of the heterogeneous blocks compound of Earth's crust, etc. Deep crustal processes are marked by heat flow and peculiarities of stress-strain state reflected in parameters of the seismic regime. The great volume of collected geological and geophysical data has revealed complex and spatially heterogeneous structures of the Earth's crust and upper mantle in this region [1–8]. The peculiarities of lithosphere forms the causes in changes in thickness and "disappearance" of granitic-gneissic layer beneath deep basins; all of this is still disputable [3, 8].

Significantly, to date, the methodology of seismotectonic formation and its relationship with geodynamics that are investigated are not enough. Small attention is paid to the study of the relationship of weak seismicity and the deep structure, the definition of suture form zones of the crystalline basement, and the connection on structural places, primarily in the sedimentary cover.

The European sector of the Arctic region is considered aseismic; however, the number of seismic events recorded here in the period of 1998–2017 with respect to

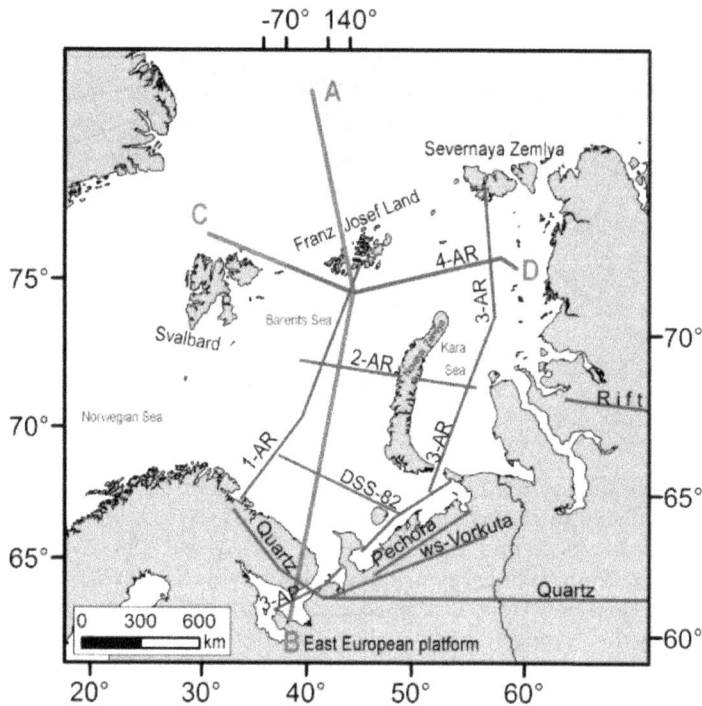

Figure 1.
Scheme showing locations of profiles A–B and C–D, geotraverses, and deep seismic profiles in water area (1-AR, 2-AR, 3-AR, 4-AR, DSS-82) and on land (Kvarts, Pechora, Rift, SW Vorkuta) in studied region.

improvements of Arctic seismic networks suggests a need for revision. The contemporary seismicity and heat flow density are indicators of geodynamic processes [9]. Joint analysis of these fields will allow a better understanding of the regional geodynamics, and such analysis is the aim of this study.

For this, a unified seismic catalog based on data from seismic networks that monitor the studied region was compiled; we generalized data of deep geological and geophysical cross sections of the crust and upper mantle along geotraverses [10–13] (**Figure 1**) and employed data on the spatial heat flow distribution [14–17]. Based on an analysis of the geological and geophysical data, we summarized the cross sections along the A–B and C–D profiles, which reflect the main structural features of the lithosphere in the region and make it possible to consider the relationship between the seismicity, heat flow density, and tectonics.

2. Relationship between the seismicity of the European Arctic and structural-tectonic elements of the lithosphere

Instrumental observations of the seismicity in the European Arctic are carried out by a number of seismic services and networks, but the Norwegian seismological center NORSAR (http://www.norsardata.no), the Arkhangelsk seismic network of N. Laverov Federal Center for Integrated Arctic Research (http://www.fdsn.org/networks/detail/AH/), and the Kola Branch of the Geophysical service of RAS (http://www.krsc.ru) make the greatest contribution.

Each seismological service has its own high-priority zones and shadow zones where earthquakes are being recorded [18]. Combining efforts in seismological

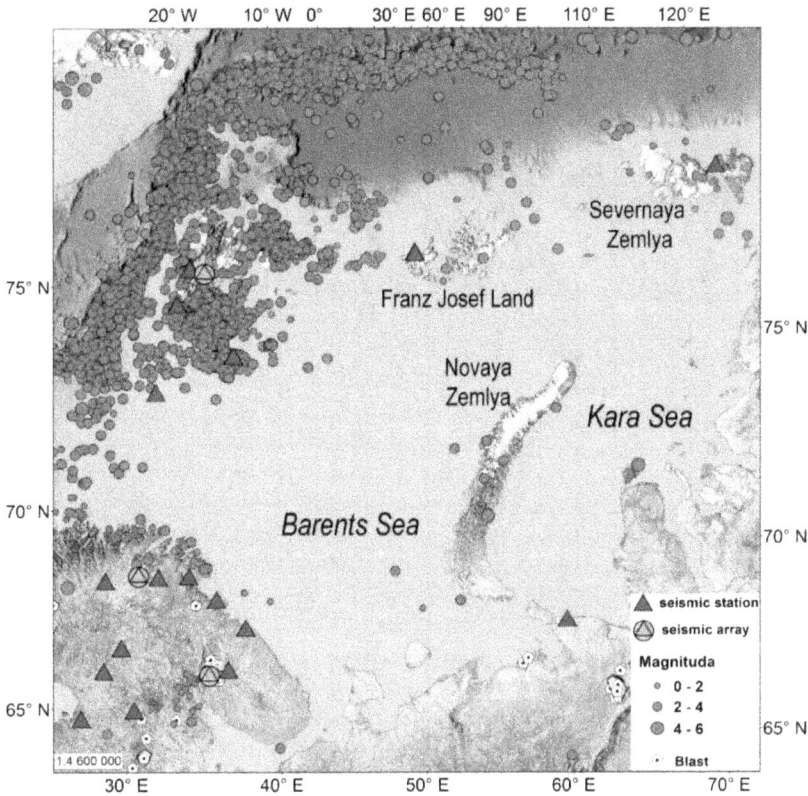

Figure 2.
Results of the unified seismic catalog which makes use of stations in Fennoscandia, Spitsbergen, Franz Josef Land, Severnaya Zemlya archipelago, north of the Russian Platform, and the Kola Peninsula. The figure covers seismic events (red dots) for 1998–2017. For Novaya Zemlya we collected information from 1986. Some additional stations which provide minor contributions to this unified catalog are not shown on the map.

monitoring of the Arctic region can help to increase the accuracy in locating epi-centers and estimating their energy. Obviously, an urgent problem is the expansion of seismic networks in the Russian sector of the Arctic region, where seismological observations are insufficient compared to foreign ones. The coverage density of seismic stations in the European Arctic is shown in **Figure 2**. The opening of several new seismic stations in the Russian Arctic recently allows to cover the European Arctic territory at large, but the number of stations is still small. The seismic sta-tions installed on the Franz Josef Land (ZFI and OMEGA) and Severnaya Zemlya (SVZ) archipelagos make a special contribution to the European Arctic monitoring, allowing to investigate the seismicity of the Gakkel Ridge (central and eastern parts) and the Arctic shelf.

Assessment of the seismic situation in the region is additionally complicated by the fact that the data in catalogs of different seismological services and networks are not unified, the quantity and quality of the initial data greatly vary, and dif-ferent processing methods have been used for them. The parameters of the same earthquakes often vary in different information sources. The seismic data were generalized to increase the quality of earthquake location in the European Arctic [9]. For each network the zones of responsibility (priority) were determined where epicentral parameters are determined with minimal errors (**Figure 2**). For example, zones of responsibility of the NORSAR network are the Mona and Knipovich ridges and Svalbard, whereas those of the Arkhangelsk seismic network are the Gakkel

Basic data							Alternative data				
Date	Origin time	Lat	Lon	Ml	Region	Data source	Origin time	Lat	Lon	Ml	Data source
08.10.2013	01.39.58.9	74.29	15.23	4.3	Mohns Ridge	NORSAR	01:40:00.0	74.44	15.30	-	ASN
08.10.2013	09:10:26.0	84.06	4.51	3.3	Gakkel Ridge	ASN	-	-	-	-	-
09.10.2013	03.32.57.0	73.17	7.31	3.0	Mohns Ridge	NORSAR	-	-	-	-	-
09.10.2013	06:13.55.0	81.42	-1.63	3.6	Knipovich region	ASN	-	-	-	-	-
16.10.2013	16:33:09.0	79.14	4.19	3.5	Knipovich region	ASN	-	-	-	-	-
18.10.2013	02:05:18.5	84.56	12.47	-	Gakkel Ridge	ASN	-	-	-	-	-
20.10.2013	16.49.53.8	72.33	2.73	3.4	Mohns Ridge	NORSAR	-	-	-	-	-
21.10.2013	19:04:49.8	86.28	49.91	3.3	Gakkel Ridge	ASN	-	-	-	-	-
22.10.2013	22.45.10.9	73.53	8.47	3.4	Mohns Ridge	NORSAR	-	-	-	-	-
23.10.2013	10.31.04.8	77.76	8.56	3.7	Knipovich region	NORSAR	10:31:08.9	77.78	8.99	3.9	ASN
23.10.2013	14:17:00.0	85.25	26.91	3.0	Gakkel Ridge	ASN	-	-	-	-	-
24.10.2013	09:51:00.8	77.92	8.50	2.2	Knipovich region	NORSAR	09:51:01.0	77.88	8.58	3.2	ASN
24.10.2013	23:46:07.0	85.01	12.02	3.4	Gakkel Ridge	ASN	-	-	-	-	-
25.10.2013	22.48.21.7	76.60	9.08	2.6	Knipovich region	NORSAR	22:48:21.0	76.77	7.79	-	ASN
25.10.2013	01:25:56.0	80.33	40.06	1.9	Franz-Victoria Graben	ASN	-	-	-	-	-

Note: ASN, Arkhangelsk seismic network; Ml, local magnitude.

Table 1.
Example of the unified seismic catalog.

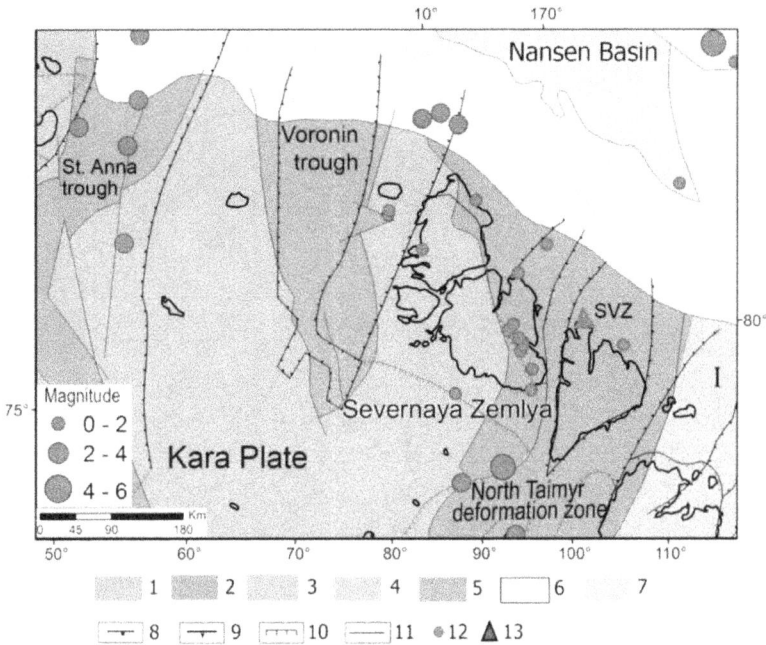

Figure 3.
Spatial distribution of seismic events in the Severnaya Zemlya archipelago region on the map of main neotectonic and geomorphological elements of the Arctic by [21]. Lithospheric plates: (1) with late Precambrian basement, (2) with late Precambrian basement that was subjected to Hercynian tectonic deformations, (3) with Grenvillian basement, (4) Neoproterozoic Taimyr accretionary belt, (5) troughs with suboceanic type crust, (6) continental slope, and (7) oceanic crust. Neotectonic faults: (8) normal faults, (9) thrusts, (10) undetermined type, (11) structures boundaries, (12) earthquakes, and (13) seismic stations. I, Taimyr accretion belt.

Ridge, Franz Josef Land, Severnaya Zemlya, and Novaya Zemlya archipelagos. The unified seismic catalog for 1998–2017 contains data on earthquakes in the European sector of the Arctic region north of 70°N, recorded by at least three seismic stations. An excerpt from the unified catalog is presented in **Table 1**.

The unified seismic catalog consists of two parts: (a) primary epicentral parameters (basic data in **Table 1**) calculated for earthquakes according to priority zones and (b) alternative versions of the earthquake parameters (alternative data in **Table 1**). There are a number of earthquakes whose parameters were calculated only by the network. According to the catalog, earthquakes in the European Arctic range from 0.9 to 6.2 in magnitude, and the representative magnitude is 2.9.

The difficulty is that the predominant numbers of earthquakes are recorded by single station only and they cannot be included in the seismic catalog (about 20% of the total number) due to the poor quality of their processing. For example, the spatial distribution of earthquakes (red circles) recorded by the SVZ station during 2017 and processed using wave forms from other seismic stations installed in the Arctic region is shown in **Figure 3**, in which the earthquake processing results recorded by the SVZ station only are presented also. Location of earthquake epicenters is reliant on the hodograph type which is absent for the central part of the Arctic Ocean; as a result, we use the NOES [19] or BARENTS [20] regional hodographs. As a result, we can determine the most likely areas of their location roughly. However, even at the first approach, these epicenters are confined to the eastern part of Gakkel Ridge, the boundary of the

Kara plate, and fall into the zone of the North Taimyr deformation associated with tectonic fault. Seismicity around Severnaya Zemlya to all appearance is consequence of rifting processes emerging in the central seismically active zone of the Laptev Sea.

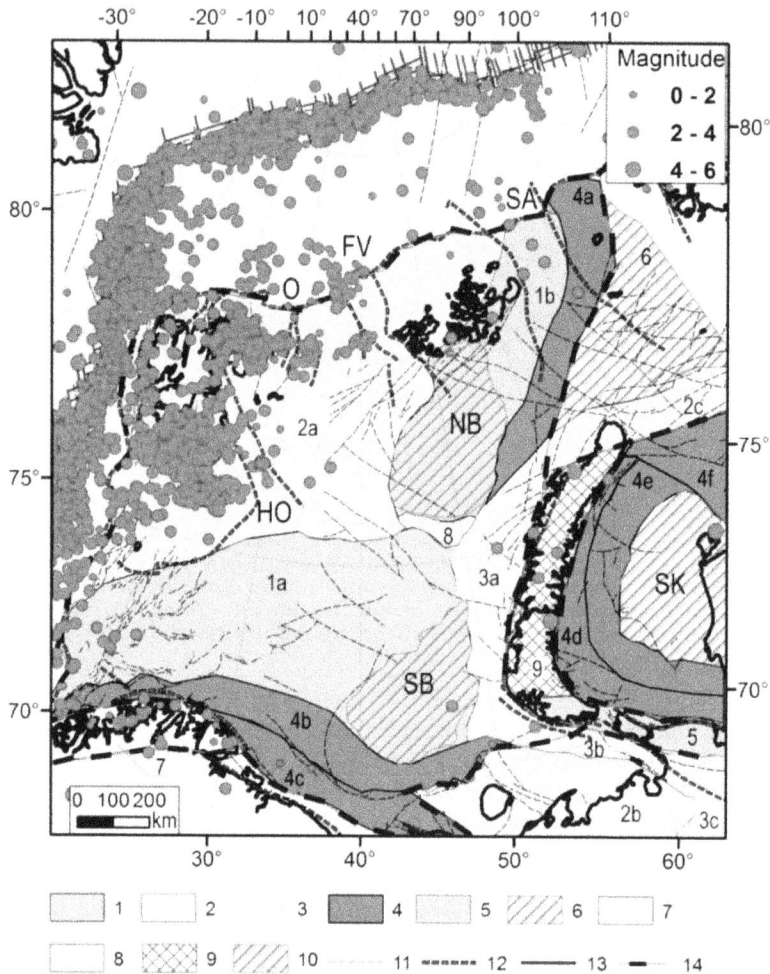

Figure 4.
Contemporary seismicity in map showing main structural-tectonic elements in Barents Sea region (with data from [13, 22]). The figure covers seismic events (red dots) for 1998–2017, and for Novaya Zemlya we collected information from 1986. Notation: SA, St. Anna trough; HO, Hipopen-Olga trench; FV, Franz-Victoria trough; O, Orly trough. (1) basins: (a) central Barents and (b) north Barents. (2) Cratonic massifs: (a) Svalbard anteclise, (b) Pechora plate, and (c) north Siberian threshold. (3) marginal troughs: (a) Sedov trough, (b) Korotaikha Basin, and (c) Kos'yu-Rogovskaya Basin. (4) slopes of deep basins: (a) east Barents step zone, (b) south Barents step zone, (c) kola monocline, (d) East Novaya Zemlya monocline, (e) East Novaya Zemlya step zone, and (f) north Siberian step zone. (5) Baikalian folding: Pai-Khoy range. (6) North Kara syneclise. (7) Caledonian folding structures of Scandinavian peninsula. (8) Luninskaya saddle. (9) early Cimmerian folding of Novaya Zemlya. (10) deep basins (SB, south Barents; NB, north Barents; SK, South Kara). (11) boundaries of near-shelf and unclassified faults. (12) largest faults, strike-slips, and thrusts. (13) active spreading center. (14) superorder structures.

Let us compare the spatial distribution of earthquakes from the unified seismic catalog and the positions of the main structural-tectonic elements in the Barents Sea region [13, 22]. By all data generalizing, we can reveal the following geodynamical peculiarities of this region (**Figure 4**):

1. Seismic activation of the arch-block ascent of Svalbard, Franz Josef Land, and the Belyi Rise was caused by tectonic stresses for which tensional and shortening morphostructures form [2, 6].

2. Extension of the continental shelf margin and its elongation in the Franz Victoria, St. Anna, and Orly toughs [2, 6], and probably isostatic compensation of rapid sedimentation at the offshore boundary, is reflected as weak seismicity within the ML magnitude range of 0.6–4.9.

3. Particular weak earthquakes were revealed in the boundaries of tectonic structures in the Central Barents Basin (Norwegian shelf) and in the Caledonian fold zone of the Scandinavian Peninsula.

4. Singular seismic event was recorded on the slopes of deep basins, namely, in the eastern Barents and southern Barents step zones:

   i. January 23, 2012, t0 = 09:52:55.0, lat 80.11, lon 72.71, ML = 2.7.

   ii. November 10, 2002, t0 = 11:04:41.7, lat 70.47, lon 49.62, ML = 2.0.

   In addition, two earthquakes were recorded in the Kola monocline:

   i. November 5, 2002, t0 = 07:31:16.22, lat 70.17, lon 34.25, ML = 1.6.

   ii. November 2, 2000, t0 = 08:14:24.61, lat 70.12, lon 36.56, ML = 1.1.

5. Seismic activity was recorded in the marginal eastern part of the Barents Sea plate, in the Novaya Zemlya fold zone, and in the Sedov Trough [23, 24]. As an example, here are two seismic events that occurred on Novaya Zemlya:

   i. October 11, 2010, t0 = 22:48:29, lat 76.18, lon 63.94, ML = 4.49.

   ii. March 4, 2014, t0 = 04:42:36, lat 74.72, lon 56.72, ML = 3.3.

We also note the event recorded in the South Barents Basin in November 11, 2009 (t0 = 04:18:20.2, lat 71.52, lon 47.06, ML = 3.2) [24]. The geological feature of the event epicenter is the big thickness of the sedimentary cover (15–20 km), which makes it unique and requires additional geophysical studies of the area.

Thus, the earthquake distribution reflects the impact of the spreading processes and transforms movements and the result of tectonic stress fields generated directly in the marginal parts of the Barents Sea plate, with singular events being recorded in its central part. The maximum cluster of earthquakes is located along the central axis of mid-ocean ridges (MOR).

3. Correlation between heat flow, seismicity, and deep structure

The time of thermal relaxation of the Earth ($\sim 1.5 \times 10^9$ years) makes it possible to consider the Earth's thermal component as constant [25]. There are two main heat sources: that supplied from the mantle ($\sim 60\%$) and that formed by radioactive decay in crustal rocks (~40%). In the sedimentary cover, the majority of radioactive elements are hosted in clay rocks, whereas intrusive bodies are the local heat sources. Based on the data from different sources [14–17], we complied a database of heat flow values. The summary data on seismicity and heat flow within the distinguished tectonic structures (cratonic and oceanic) of the studied region are presented in **Table 2**.

The seismic activity correlates with heat flow values in middle ocean ridge (MOR) areas. In the northern East European Craton, there was no clear relationship between these parameters, except for the North Barents Rise. Let us consider

Structures		Earthquakes			Average heat flow, mW/m^2
Superorder structures	First and second orders	Number	ML$_{max}$	ML$_{av}$	
Barents plate	Central Barents Basin (1a)	23	3.6	2.48	60–70
	North Barents Basin (1b)	3	2.7	2.4	60–80
	North Barents Rise (2a) Orly trough (exclusion)	1758	5.9	2.5	60–80, 100–300
	Sedov trough (3a)	2	2.3	2.25	50–80
	East Barents step zone (4a)	1	2.7	2.7	70
	Luninskaya saddle (8)	-	-	-	70
	South Barents step zone (4b)	9	3.7	2.37	60–70
	Kola monocline (4c)	2	1.6	1.35	50–60
Timan-Pechora plate	Pechora plate (2b)	-	-	-	40–50
	Korotaikha Basin (3b)	-	-	-	40
	Kos'yu-Rogovskaya Basin (3c)	-	-	-	30–40
	Timan Range (5a)	-	-	-	50
West Siberian plate	East Novaya Zemlya monocline (4d)	-	-	-	60
	East Novaya Zemlya step zone (4e)	-	-	-	60
	North Siberian step zone (4f)	-	-	-	60
Novaya Zemlya microplate	Early Cimmerian folding of Novaya Zemlya (9)	5	4.5	3.24	60
	Pai-Khoy Range (5b)	-	-	-	60
North Kara plate	North Siberian threshold (2c)	-	-	-	60–70
	North Kara syneclise (6)	-	-	-	70
Baltic Shield	Caledonian folding structures of Scandinavian Peninsula (7)	33	2.8	1.9	40–50
Eurasian and North Atlantic Basins	Nansen Basin	135	4.3	2.5	60–80
	MOR	3224	6.6	2.83	>100

Table 2.
Seismicity parameters and heat flow in distinguished tectonic structures of studied region of the Arctic.

the distribution of these parameters in more detail, with knowledge about the structure of the lithosphere along the composite geological and geophysical cross sections (**Figure 1**).

Profile A–B (**Figure 5**) crosses such morphostructures as the MOR (Gakkel Ridge), the abyssal plain (Nansen Basin), the Barents Sea shelf, the eastern Baltic Shield, the White Sea shelf, and the continental rise of the East European Craton. To construct the model for the lithosphere structure on P-wave velocities, we used data from deep geological and geophysical cross sections along such profiles as Kvarts, 1-AR, 2-AR, 3-AR, 4-AR, DSS-82, etc. [10–13, 26].

Profile C–D (**Figure 6**) crosses such morphostructures as the Svalbard anteclise, North Barents Basin, North Kara syneclise, and the Taimyr-Severnaya Zemlya fold system. The majority of it is overlapped by the 4-AR deep seismic profile (seismic reflection CMP method) [2, 11, 13].

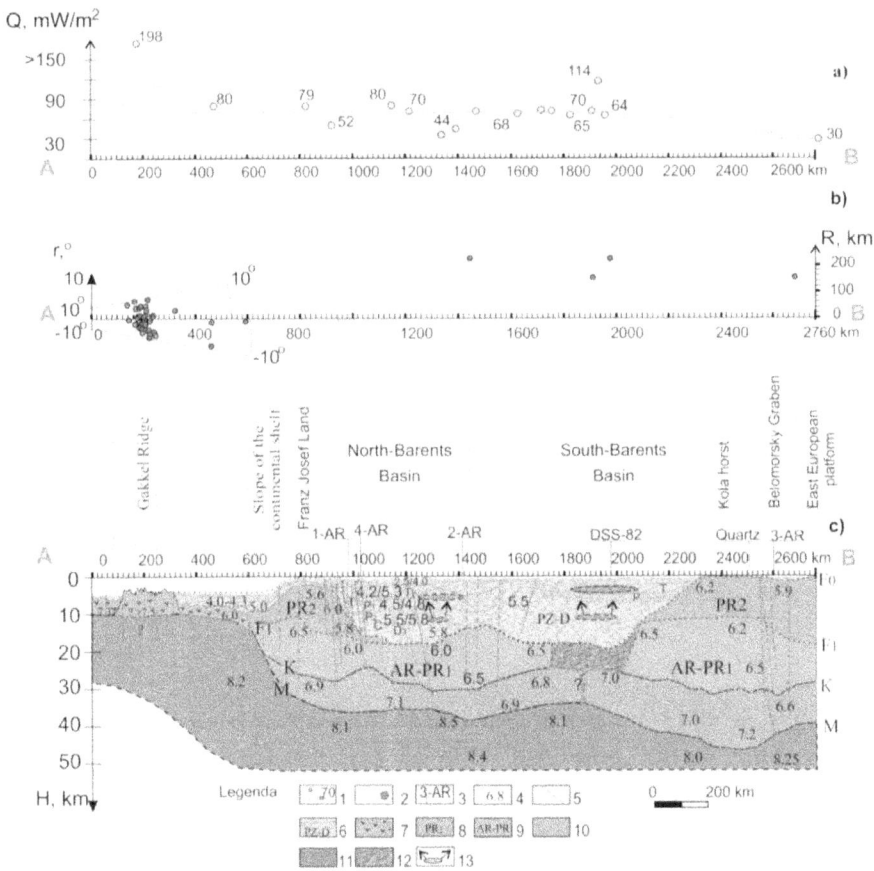

Figure 5.
Distribution of heat flow (I) and seismicity (II) and geological and geophysical cross section along profile A–B (III) (with data from [10, 13, 26]). Notation: M, Moho; K, middle boundary in crust; Fo, top of upper Proterozoic basement; and F1, top of Archean-Proterozoic crust (PR1-AR). Arbitrary notes: (1) heat flow values (including averaged ones) along profile, mW/m², (2) epicenters of earthquakes in 1995–2015, (3) crossing points of geotraverses, (4) P-wave velocities, (5) faults, (6) sedimentary cover and its age, (7) acoustic basement of oceanic crust, (8) upper Proterozoic basement (PR2), (9) upper sialic part of consolidated crust (PR1-AR), (10) basite part of consolidated crust, (11) upper mantle, (12) basite massif, and (13) fluid-saturated decompacted zones in sedimentary cover where hydrocarbon generation is possible.

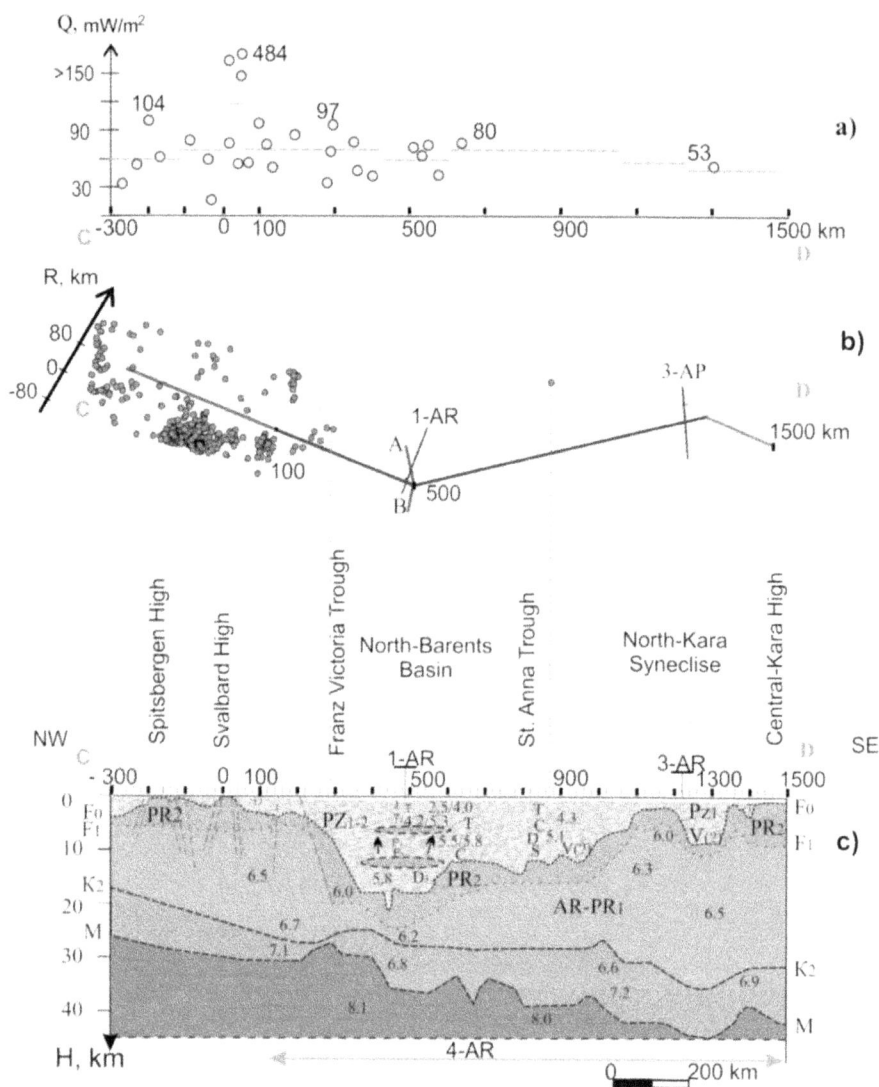

Figure 6.
Distribution of heat flow (I) and seismicity (II) and geological and geophysical cross section along profile C-D (III) (with data from [10, 13, 26]). Notations are shown in **Figure 5.**

The heat flow values and earthquake epicenters are drawn along the composite lithosphere cross sections. Below are the results of comparison of the geological and geophysical fields.

3.1 Profile A–B

The oceanic lithosphere is sharply distinguishable from the continental crust, and the oceanic Moho is located at a depth of 12–13 km (**Figure 5**). The seismic velocities in the upper oceanic crust are from 4.5 to 6 km/s, whereas they are from 6.8 to 7.3 km/s in the lower part. The rift valley of the Gakkel Ridge is formed by rocks from the oceanic basement, which supposedly had velocities of more than 7.5 km/s [26]. In the sedimentary cover above the basement of the oceanic crust, we can distinguish several stratigraphic complexes, whose thicknesses increase toward the Barents-Kara continental margin. In the Nansen Basin, the Moho was acoustically detected at 10–12 km depth [26].

Seismic activity has been recorded in the zone where the continental and oceanic lithosphere joins. Single seismic events in this area are supposedly caused by the removal of sedimentary masses from the continent [2] or by transform fault activity. The lithosphere is the continental type. The surface of the mantle (P-wave velocities from 8.0 to 8.5 km/s) is at 34–36 km in the Barents Sea Basin, 35–40 km in the Kola monocline, and 44–46 km in the Baltic Shield. The consolidated crust can be roughly subdivided into two layers. The upper one has velocities of 5.6–6.5 km/s and the lower 6.6–7.2 km/s. The thickness of the upper layer of the consolidated crust changes from 8 km in the basin to 15–25 km in the areas of the Voronin, Albanov, and Fedynsky rises, as well as beneath the Baltic Shield [8]. The thickness of the basite part of the consolidated crust in this area ranges from 10 to 20 km in the zone of the rise, whereas it thins to 5–16 km in the sinking zone [10, 11].

For the oceanic lithosphere with thinned crust, heat flow increases to 200 mW/m^2, and seismic activity is higher, especially in the Gakkel Ridge area. In the Nansen Basin, where single earthquakes have been recorded closer to the transform zones, the heat flow values are 98 mW/m^2. Toward the ledge of the continental shelf, the average heat flow decreases to 70 mW/m^2.

Based on the seismic data, the upper layer of the consolidated crust beneath the South Barents Basin contains local velocity inhomogeneities. It is assumed that the upper crust contains abundant plateau basalts and is close to oceanic crust in its physical properties. Such thinning and transformation of the continental crust, coupled with its sinking, were probably caused by phase transitions of rocks [1, 27]. **Figures 5** and **6** schematically show fluid-saturated decompaction zones in the sedimentary cover, where subsequent generation of hydrocarbons is possible. In the North Barents Basin, the lower crustal layer contains high-velocity inhomogeneities with values of 7.1 km/s. According to [1, 28], their compactions were the result of gradual metamorphic transition of gabbroids to eclogite.

In the Barents Sea Basin, singular earthquakes have been recorded in zones where rock transformation takes place (**Figure 4**). These areas are remarkable for higher heat flow values (60–80 mW/m^2). The last zone along profile A–B is a thick mantle-crustal structure; it is thermally cold (heat flow values from 30 to 50 mW/m^2) and has a thin, up to wedging, sedimentary layer in the southwestern part of the continental rise of the East European Craton. It can be assumed that this structure limits shortening from the Middle Arctic Ridge and tectonic deformations from the fold units of the Polar Urals, Novaya Zemlya, Taimyr Peninsula, and Caledonides of the North Atlantic, which is manifested as single relatively weak earthquakes at the boundaries of large tectonic structures [29].

3.2 Profile C–D

High seismic activity has been recorded in the area of collisional dislocations at the Svalbard plate margin (**Figure 6**). Heat flow values of about 80 mW/m^2 exceed the mean ones for the Barents Sea Rise, and in the area of Orly Trough, they reach peak values of ≈500 mW/m^2. The slow and gentle downwarping of this part of the Barents Sea plate to an almost horizontal plane resulted in the formation of the North Kara Basin [6].

Seismic data suggest that the crust is continental type, which is supported by low seismic wave velocities (5.6–6.0 km/s) in the granitic-gneissic layer. According to [8, 11], thinning of the crust and the influence of small flows of deep mantle fluids (compared to those that affected the South Barents Basin) gave rise to slow eclogitization in the lower crust. This is supported by the mean heat flow (about 70 mW/m^2), with an anomalous increase up to 97 mW/m^2. In the eastern part of profile C–D, there are only singular instances of heat flow data, and these are 50 mW/m^2 on average.

Seismic activity manifests itself in most of profile C–D, although it decreases moving from the junction zone between the Svalbard plate and the MOR. This may be related in part with the absence of permanently operating seismic stations in the Kara Sea region, because the model from [18] suggests that there should be weak seismicity at the junction of the Barents Sea and North Kara plates. The East Barents step zone (**Figure 4**) has the only seismic event recorded (lat 80.11, long 72.71, ML = 2.7), supposedly above the domain of high-velocity inhomogeneities in the lower and upper crusts (**Figure 6**).

4. Conclusion

Comparison of the geological and geophysical fields makes it possible to combine together different indicators of geodynamic processes. We have revealed the relationship between seismicity and deep structure, as well as the correspondence of seismicity to both the most geodynamically active structures and zones of concentrated tectonic stresses. According to the compiled unified catalog, the most geodynamically active area in the Barents Sea region is the Svalbard anticline, where the greatest concentration of tectonic stresses is observed on the MOR side and zones with higher values of heat flow are distinguished (70–80 mW/m^2 on average).

The recorded single earthquakes in the basins and troughs reflect manifestations of recent tectonic activity in the region, probably as a result of commonly developed high-velocity inhomogeneities in the lower and upper crusts or accumulation and release of stresses in weakened zones. This conclusion is, however, quite conditional because of a small time interval for which seismic data on the Barents Sea region are available. Seismological monitoring is necessary in the future to verify or refute such conclusions.

Manifestations of single earthquakes should be marked as promising areas for searching and prospecting for hydrocarbon fields within the limits of the South and North Barents basins, St. Anna Trough, and western Barents Sea Basin. Our studies supplement existing data on the structure and tectonics of the Barents Sea region.

Acknowledgements

This work is supported by the Russian Foundation for Basic Research, Grant No. 18–05–70018 "Geodynamic situation of oil and gas-bearing basins of Barents-Kara region according to the recent seismotectonic data."

We are grateful to Khutorskoi, DSc, MD (Geological Institute of the Russian Academy of Sciences, Moscow), for his recommendations on geodynamic studies of the Arctic. We also acknowledge NORSAR (Norway) and the Kola Branch of the United Geophysical Survey, Russian Academy of Sciences, for the joint work in compiling the unified seismic catalog for the European sector of the Arctic.

Conflict of interest

The authors declare no conflict of interest.

Author details

Evgeniy Rogozhin[1], Galina Antonovskaya[2*], Irina Basakina[2] and
Natalia Kapustyan[1,2]

1 Institute of Physics of the Earth of the Russian Academy of Sciences, Moscow,
Russian Federation

2 N. Laverov Federal Center for Integrated Arctic Research, Arkhangelsk,
Russian Federation

*Address all correspondence to: essm.ras@gmail.com

IntechOpen

References

[1] Artyushkov EV. Mechanism of formation of superdeep sedimentary basins: Lithospheric stretching or eclogitization? Geology and Geophysics. 2010;**51**:1304-1313

[2] Verba ML. Present-day bilateral extension of the crust in the Barents–Kara region and its role when evaluating the petroleum-bearing potential (Vol. 2). In: Petroleum Geology: Theory and Practice [Internet]. 2007. Available from: http://www.ngtp.ru/rub/2007/026.html [Accessed: 16-07-2018] (in Russian)

[3] Kashubin SN, Pavlenkova NI, Petrov OV, Mil'shtein ED, Shokal'skii SP, YuM E. Crustal types in the Circumpolar Arctic. Reg. Geol. Metallog. 2013;**55**:5-20

[4] Laverov NP, Lobkovsky LI, Kononov MV, Dobretsov NL, Vernikovsky VA, Sokolov SD, et al. A geodynamic model of the evolution of the Arctic Basin and adjacent territories in the Mesozoic and Cenozoic and the outer limit of the Russian continental shelf. Geotectonics. 2013;**47**:1-30

[5] Sim LA, Zhirov DV, Marinin AV. Strain state reconstruction for the eastern Baltic Shield. Geodinamika I Tektonofizika. 2011;**3**:219-243

[6] Sorokhtin NO, Nikiforov SL, Koshel' SM, Kozlov NE. Geodynamic evolution and morphostructural analysis of the western sector of the Russian Arctic shelf. Vestn. Murmansk. Gos. Tekh. Univ. 2016;**19**(1):123-137

[7] Yudakhin FN, Shchukin YK, Makarov VI. Deep Structure and Contemporary Geodynamic Processes in the Lithosphere of the East European Craton. Yekaterinburg: Ural. Otd. Ross. Akad. Nauk; 2003 [in Russian]

[8] Artyushkov EV, Belyaev IV, Kazanin GS, Pavlov SP, Chekhovich PA, Shkarubo SI. Formation mechanisms of ultradeep sedimentary basins: The North Barents basin. Petroleum potential implications. Russian Geology and Geophysics. 2014;**55**:649-667

[9] Antonovskaya GN, Basakina IM, Konechnaya YV. Distribution of seismicity and heat flow anomalies in the Barents Sea region. Geotectonics. 2018;**52**(1):45-55. DOI: 10.1134/S001685211801003X

[10] Pavlenkova NI, Pavlenkova GA. Structure of the Earth's Crust and Upper Mantle of Northern Eurasia Based on Seismic Profiling Data from Nuclear Blasts. Vol. 10. Moscow: GEOKART; 2014 [in Russian]

[11] Sakulina TS, Verba ML, Ivanova NM, Krupnova NA, Belyaev IV. Deep structure of the Northern Barents–Kara region along the 4-AR reference profile (Taimyr Peninsula–Franz Josef Land). In: Models of the Crust and Upper Mantle from Results of Deep Seismic Profiling: Proceedings of the International Workshop for Science and Practice. St. Petersburg: VSEGEI; 2007. pp. 197-200

[12] Sakulina TS, Roslov YV, Pavlenkova GA. Methods and results of processing of complex seismic investigations on the 2-AR profile (Barents–Kara Shelf). Izvestiya, Physics of the Solid Earth. 2009;**45**:231-238

[13] Spencer AM, Embry AF, Gautier DL, Stupakova AV, Sorensen K. Arctic Petroleum Geology. Vol. 35. London: Geological Society of London, Memoirs; 2011. DOI: 10.1144/M35.21

[14] Khutorskoi MD, Akhmedzyanov VR, Ermakov AV, Leonov YG, Podgornykh LV, Polyak BG, et al.

Geothermy of the Arctic Seas. Moscow: GEOS; 2013 [in Russian]

[15] Khutorskoi MD, Leonov YG, Ermakov AV, Akhmedzyanov VR. Abnormal heat flow and the trough's nature in the Northern Svalbard Plate. Doklady Earth Sciences. 2009;**424**:29-35

[16] Davies JH, Davies DR. Earth's surface heat flux. Solid Earth. 2010;**1**:5-24

[17] The Global Heat Flow Database of the International Heat Flow Commission, University of North Dakota [Internet]. Available from: http://www.heatflow.und.edu/data.html [Accessed: 16-07-2018]

[18] Antonovskaya G, Konechnaya Y, Kremenetskaya E, Asming V, Kvaerna T, Schweitzer J, et al. Enhanced earthquake monitoring in the European Arctic. Polar Science. 2015;**9**:158-167

[19] Morozov AN, Vaganova NV, Ivanova EV, Konechnaya YV, Fedorenko IV, Mikhaylova YA. New data about small-magnitude earthquakes of the ultraslow-spreading Gakkel Ridge, Arctic Ocean. Journal of Geodynamics. 2016;**93**:31-41

[20] Kremenetskaya E, Asming V, Ringdal F. Seismic location calibration of the European Arctic. Pure and Applied Geophysics. 2001;**158**(1-2):117-128

[21] Vernikovsky VA, Dobretsov NL, Metelkin DV, Matushkin NY, Koulakov IY. Concerning tectonics and the tectonic evolution of the Arctic. Russian Geology and Geophysics. 2013;**54**:838-858

[22] Stupakova AV. Structure and petroleum-bearing potential of the Barents–Kara Shelf and adjacent areas. Geol. Nefti Gaza. 2011;**6**:99-115

[23] Gibbons SJ, Antonovskaya G, Asming V, Konechnaya YV, Kremenetskaya E, Kvaerna T, et al. The 11 October 2010 Novaya Zemlya earthquake: Implications for velocity models and regional event location. Bulletin of the Seismological Society of America. 2016;**106**. DOI: 10.1785/0120150302

[24] Morozov AV, Asming VE, Vaganova NV, Konechnaya YV, Mikhaylova YA, Evtyugina ZA. Seismicity of the Novaya Zemlya archipelago: Relocated event catalog from 1974 to 2014. Journal of Seismology. 2017;**21**:1439-1466. DOI: 10.1007/s10950-017-9676-y

[25] Zharkov VN. Inner Structure of the Earth and Planets. Moscow: Nauka; 1983 (in Russian)

[26] Poselov VA, Pavlenkin AD, YuE P, Kaminskii VD, Murzin PP, MYu S. Structure of the Arctic Basin lithosphere from seismic data with respect to the problem of the outer boundary of Russian continental shelf zone. Razved. Okhr. Nedr. 2000;**12**:48-54

[27] Pavlenkova NI, Kashubin SN, Pavlenkova GA. The Earth's crust of the deep platform basins in the Northern Eurasia and their origin. Izvestiya Physics of the Solid Earth. 2016;**52**:770-784

[28] Artyushkov EV. Mechanism of the Barents trough formation. Geologiya i Geofizika. 2005;**46**:698-711

[29] Yudakhin FN. Lithosphere and hydrosphere of the Northern European Part of Russia: Environmental Problems. Yekaterinburg: Ural. Otd. Ross. Akad. Nauk; 2001 [in Russian]

Chapter 5

The Effect of Space Weather on Human Body at the Spitsbergen Archipelago

Natalia K. Belisheva

Abstract

The study of the effects of the space weather on the human body was carried out at the Spitsbergen archipelago. A geophysical feature of the arch. Spitsbergen is its location in the cusp region—a kind of funnel on the dayside of the magnetosphere, where phenomena of space weather most express. Diverse radiation (from ULF to VHF) and waves in the field of polar cusp, covering the entire range of the body rhythms, give credit for studying the effects of space weather in the field of polar cusp. Assessment of the relationship between the dynamics of the monthly morbidity in Russian settlements and indicators of space weather revealed that, practically, all forms of morbidity are associated with solar activity and with the local geomagnetic activity in the polar cusp. A difference in correlations between the monthly incidence of residents in the Barentsburg and geocosmic agents during the polar day and the polar night was found. The links between the incidences of the population and the peculiarities of space weather will make it possible to develop prognoses of the morbidity for preventive measures aimed at increasing human health in high latitudes.

Keywords: space weather, morbidity, Spitsbergen archipelago,
polar day and polar night

1. Introduction

The Spitsbergen archipelago is located in the Arctic Ocean, between 76° 26' and 80° 50' north latitude and 10 and 32° east longitude. A geophysical feature of the arch. Spitsbergen is its location in the cusp region [1]—a kind of funnel on the dayside of the magnetosphere with near zero magnetic field magnitude, where, under certain conditions, the solar wind (CW) can burst through powerful plasma jets (**Figure 1**, [2]). The open field lines of the cusp is connected with those of the interplanetary magnetic field (IMF), which allows the shocked solar wind plasma of the magnetosheath to enter the magnetosphere and to penetrate the ionosphere [3].

The Earth's magnetosphere is a highly dynamic structure that responds dramatically to solar variations [4, 5], especially in the cusp region [6]. The upper atmosphere at high latitudes, associated with cusp, is also called the "Earth's window to outer space." Through various electrodynamic coupling processes as well as through direct transfer of particles, many geophysical effects displayed that there are direct manifestations of phenomena occurring in the deep space. In the polar cusps, the

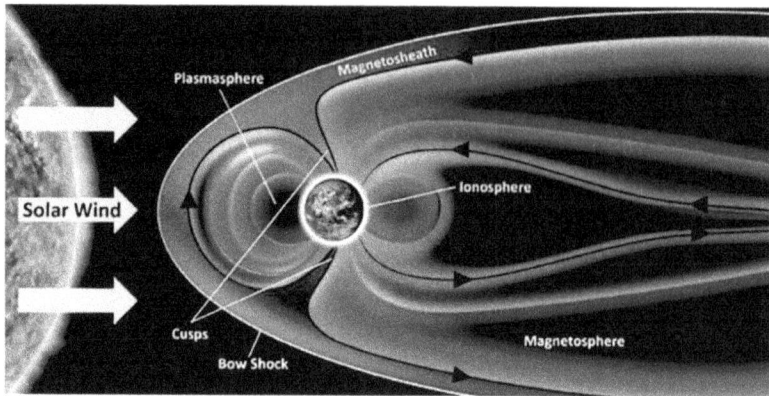

Figure 1.
Earth's protective shield: **magnetosphere** *is that area of space, around a planet, that is controlled by the planet's magnetic field, whose shape is the direct result of being blasted by solar wind; a supersonic shock wave is created sunward of earth called the* **Bow shock**; *the* **magnetosheath** *is the region of space between the magnetopause (the outer boundary of Earth's confined geomagnetic field) and the Bow shock; the* **plasmasphere**, *or inner magnetosphere, is a region of the Earth's magnetosphere consisting of low energy (cool) plasma; the* **ionosphere** *is the ionized part of Earth's upper atmosphere; the polar* **cusps** *are funnel-shaped regions in the frontal part of the magnetopause at geomagnetic latitudes of ~75°.*

solar wind plasma has also direct access to the upper atmosphere. The polar regions are thus of extreme importance when it comes to understanding the physical processes in the near space and their effect on our environment" [6].

In the cusp areas, the impacts of the solar wind (SW) on the Earth's magnetosphere manifest most strongly, and multiple phenomena originating as consequences of such interactions are referred to as space weather. It can be truly said that space weather affects everybody, either directly or indirectly. Space weather is defined by the U.S. National Space Weather Program (NSWP) as "conditions on the Sun and in the solar wind, magnetosphere, ionosphere, and thermosphere that can influence the performance and reliability of space-borne and ground-based technological systems and can endanger human life or health" [7, 8].

2. Magnetosphere-ionosphere emissions and waves in the polar regions

Space weather begins at the sun. The sun exhibits an 11-year cycle of sunspots that are visible manifestations of increased solar magnetic field. Certain larger flares produce solar radio bursts of broadband noise from 10 MHz to 10 GHz that may directly affect GPS receivers on the dayside of the earth. Terrestrial effects are the result of three general types of conditions on the Sun: eruptive flares, disappearing filaments, and coronal holes facing Earth [9], on which the nature of magnetosphere-ionosphere interactions depends. The magnetosphere and the ionosphere of the Earth are sources of electromagnetic oscillations and waves, many of which are detected in the form of radiation outside the region of generation, in particular, on the surface of the Earth. The electromagnetic radiation range of the magnetosphere and ionosphere overlaps in frequency by many orders of magnitude—from the lowest frequencies of magnetohydrodynamic (MHD) waves ($f \sim 5$–10^{-3} Hz) to X-rays of energetic electrons in the upper atmosphere ($f \sim 10^{18}$ Hz) [10]. The complexity and diversity of physical phenomena associated with solar activity and transmitted to earth through solar-terrestrial connections make the issue of identifying bioeffective agents in the space weather phenomenon nontrivial and rather complicated. Some of the cosmophysical phenomena, as attributes of space weather, are most pronounced and specific for high latitudes and for the polar cusp [11–14].

3. The effect of space weather on human body at the Spitsbergen archipelago

3.1 Material and methods

The unique data characterizing morbidity of the residents in the Russian settlements of the Barentsburg (1985–1993), including the females, were used in the study. The statistics on the complications about pregnancy and the postpartum period in women, who lived in the archipelago during the time of the former USSR, provide invaluable information that allows assessing the effect of space weather associated with the polar cap and the polar cusp on pregnant women. Today, such research is extremely difficult, because the residence of pregnant women in the Spitsbergen archipelago is undesirable.

The monthly statistical reports on the morbidity structure in the Barentsburg mine hospital (1985–1993) were basis for analysis [15]. All data of morbidity were normalized on 1000 people of residents in the Barentsburg. The average number of inhabitants in each Russian settlement (Barentsburg and Pyramid) was about 1000, where one third were women. The average monthly data characterizing the CA were selected in the National Geophysical Data Center (NGDC): Solar Data Services (http://www.ngdc.noaa.gov/stp/SOLAR/ftp: sunspotnumber.html); intensity of the secondary cosmic rays (CR) was estimated by neutron count rate (ground station of the neutron monitor of the PGI KSC Russian Academy of Sciences in the Apatity and in the Barentsburg). Statistical data analysis was performed using the software Statistica 10.0 and the graphing was carried out using the software package ORIGIN50.

3.2 Results and discussion

3.2.1 Monthly morbidity

The bioefficiency of geocosmic agents is manifested in synchronous dynamics of the functional state of resident's organism in the high latitudes [16–21] or in the coherency of morbidity dynamics of the population in the Arctic territories [22] with variations of the geocosmic agents on the time scales with different resolution (day, month, and year).

The coefficients of cross-correlations between the monthly (01.01.1985–31.12.1989) values on the curves, smoothed by 5 points, of the morbidity and the solar radio flux f10.7_index are demonstrated in **Table 1**.

The synchronism of the incidence diseases follows from the cross-correlation coefficients shown in **Table 1**, where one can see that the monthly values of the incidence of the mental disorders (MD) have significant correlation coefficients with injuries and poisonings (IP) and with f10.7-index. However, IP correlates with other diseases (**Table 1**): with DEA), with DAAV, with IFGO, with ISST, and with the fluxes of solar radio emission (f10.7-index).

One can see certain concordance between the curves of the average monthly angular parameters of the solar wind (sigma-phi-V, deg.,), the monthly dynamics of incidence of the mental disorders (MD), the injuries, and poisoning (IP) in **Figure 2A**. Coefficient correlations between sigma-phi-V and the MD, sigma-phi-V, and IP are r = 0.32, r = 0.44, and p < 0.05, respectively. In this case, the MD and the

	MD	DEA	DAAV	IFGO	ISST	DMSCT	IP	f10.7_index
MD	1.00							
DEA	0.72	1.00						
DAAV	0.39	0.43	1.00					
IFGO	0.25	0.41	0.22	1.00				
ISST	0.33	0.45	0.28	0.34	1.00			
DMSCT	0.61	0.64	0.83	0.48	0.47	1.00		
IP	0.50	0.53	0.77	0.51	0.50	0.86	1.00	
f10.7_index	0.46	0.70	0.69	0.67	0.56	0.83	0.87	1.00

The incidence of the mental disorders (MD); the diseases of the eye and its appendages (DEA); the diseases of arteries, arterioles and veins (DAAV); the incidence of the inflammatory processes of the female pelvic organs and other diseases of the female genital organs (IFGO); the infections of the skin and subcutaneous tissue (ISST); diseases of the musculoskeletal system and connective tissue (DMSSCT); the injuries and poisoning on the way to and from work (IP) and the solar radio flux with wavelength 10.7 cm (f10.7-index). Significant correlations are marked by red color.

Table 1.
Coefficients of cross-correlations (p < 0.05) between the monthly (01.01.1985–31.12.1989) values of the morbidity.

IP diseases are not only interconnected by connection with the solar radio emission (**Table 1**, f10.7-index), but also with the parameters of the solar wind (sigma-phi-V, deg). This suggests that the solar wind could generate such conditions in the cusp area, when the physical agents might affect the mental state, and through it, the predisposition to the appearance of the injury.

In **Figure 2B**, one can see concordance between curves of average monthly variations of the solar radio flux at 10.7 cm, dynamics of monthly diseases of arteries, arterioles, and veins and (DAAV), average monthly values of the sigma-theta-V, deg. and average monthly values of the Pc (N)-index. Correlation coefficients between f-10.7-index, sigma-theta-V, deg., Pc (N), and DAAV are r = 0.40; r = 0.29; r = 0.27; and p < 0.05, respectively. The positive relationship between the incidence of DAAV, the f-10.7-index, and PC(N) means that with increasing solar activity and associated geomagnetic disturbances, the morbidity of DAAV also increases. The connection between PC and the DAAV demonstrates the effect of the space weather on the vascular system of human organism.

Figure 3A shows the connection between the dynamics of monthly pregnancy complications (IFGO), the parameter of space weather (hydrodynamic pressure of the solar wind), and the ap-index reflecting the local geomagnetic activity. The connection between the dynamics of monthly inflammatory processes of the female pelvic

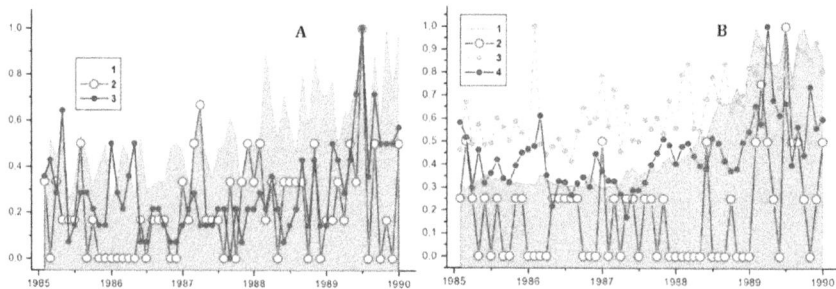

Figure 2.
Coherent dynamics of the average monthly values of the parameters of geocosmic agents and the monthly values of morbidity. A. Parameter of solar wind "sigma-phi-V, deg" (1, graph area—cyan), the incidence of the mental MD (2), incidence of the injuries and poisoning on the way to and from work, IP (3). B. Solar radio emission with wavelength 10.7 cm (1, graph area—cyan), incidence of the diseases of arteries, arterioles, and veins (2), sigma-theta-V, deg. (3), pc (N) index (4). X axis: The months of the year from January 1985 to December 1989; Y axis: the normalized values of the all parameters.

organs and other diseases of the female genital organs (IFGO), the F10.7-index, and PC(N) are shown in **Figure 3B**. Correlation coefficients between IFGO, flow pressure, and ap-index are r = 0.34; r = 0.29, respectively, p < 0.05. Correlation coefficients between IFGO, F10.7-index, and PC(N) are −r = 0.34; r = 0.29, respectively, p < 0.05.

One can again remark that morbidity, even specific such as diseases of the female genital system, is associated with solar and geomagnetic activity, expressed by the ground indicators of local geomagnetic storm conditions PC(N), ap-index), and the agents in the near Earth space (F10.7-index, variations of the angle velocity of solar wind—sigma-phi-V, sigma-teta-V, deg., flow pressure). The fluctuations of the monthly values of morbidity of the somatic diseases, the mental disorders, and the frequency of injuries and poisoning, as well as the coherency of the diseases among themselves and with the space weather indicators suggest that space weather controls the state of the human body in Svalbard.

3.2.2 Association of the monthly morbidity with space weather agents in the polar day and in the polar night

The source of physical phenomena, some of them could have a pronounced bioefficiency, is the magnetosphere-ionosphere interaction, reflecting the interaction of the solar plasma with the earth's magnetosphere in the polar cusp region. Since the properties of the ionosphere are largely determined by Solar X-*rays* and UV *radiation* as well as fluctuations in the concentration of particles associated with magnetic disturbances, the properties of the ionosphere in the polar cusp region should differ during the polar day and in the polar night.

Ionospheric differences during the polar day and the polar night are also confirmed by differences in the electrical current systems in the summer season and in the winter due to current vortex, which is most noticeable in the summer season [23]. The total electron content (TEC) exhibits significant spatial and temporal variations, when the minimum level of TEC observed in the high latitude of the northern hemisphere in the mid polar night (December) and the maximum level— in the mid polar day [24]. A characteristic feature of geomagnetic disturbances in all hours is the presence of pulsations with large amplitudes and periods of several minutes. And some of them practically disappear during the polar night [14, 25–29].

To appreciate the significance of the space weather agents (geocosmic agents) affecting the human health in the polar days and in the polar night, the monthly data sets of the morbidity in the settlement of Barentsburg were sorted in two

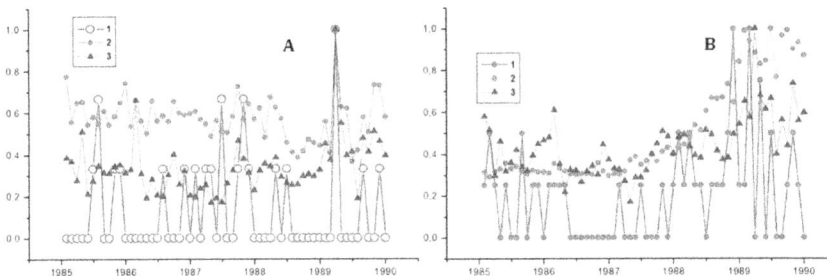

Figure 3.
Coherency dynamics of the monthly values of morbidity and the monthly average values of the parameters of geocosmic agents. A. Incidences of the complications of pregnancy and the postpartum period, CPP (1); flow pressure of the solar wind, (nPa) (2); and ap-index (3). B. Incidences of the inflammatory processes of the female pelvic organs and other diseases of the female genital organs, IFGO(1), the solar radio emission with wavelength 10.7 cm (2), PC(N)-index. X axis: the months of the year from January 1985 to December 1989; Y axis: the normalized values of the all parameters.

groups. In the first group was included the monthly values of morbidity in the polar day (from March to September, n = 35) and in the second group—the monthly values of morbidity in the polar night (from October to February, n = 25). This sorting was performed due to the duration of the dark time (122 days) from 21 October to 20 February at 80 degrees north latitude [30]. Significant differences between the incidence of the population during the polar day and the polar night, as well as differences in the values of geophysical indicators, have been estimated by using the nonparametric (the Mann-Whitney U test, Kolmogorov-Smirnov criterion) and the parametric T-criterion.

It turned out that the monthly values of incidence during the polar day and night significantly differ only in cases of intestinal infections (yersiniosis) and the inflammatory processes of the female pelvic organs and other diseases of the female genital organs (IFGO). During the polar day and the polar night, incidences of intestinal infections were 0.05 ± 0.21 and 0.25 ± 0.49, respectively, $p < 0.05$; incidences of IFGO were 1.89 ± 2.58 and 3.70 ± 3.62 in the polar day and in the polar night (according to the Mann-Whitney U-test T-criterion). The geophysical indices differed only in the monthly average values of atmospheric pressure (992.36 ± 4.01 and 987.58 ± 7.70, $p < 0.005$, mb), in the Bulk flow latitude (2, 24 ± 0.67 and 1.16 ± 0.93, $p < 0.001$, degrees), in the DST index (-16.07 ± 12.79 and $- 22.16 \pm 8.91$, nT, $p < 0.025$), and in the PC (N) index (0.96 ± 0.35 and 1.14 ± 0.24, $p < 0.005$), respectively, in the polar day and in the polar night. That is, in fact, the incidence rate on the polar day and on the polar night, with a few exceptions, just as the monthly average of geophysical agents, with the exception of 2 indices characterizing geomagnetic activity, does not differ.

However, when correlations between the monthly values of morbidity and the monthly average values of geophysical agents corresponding to the polar day and to the polar night periods were compared, it turned out that there are large differences between them. These differences indicate that during the polar day and during the polar night, the roles of similar geophysical agents are different.

One can see above (**Table 1**) that the monthly values of the incidences of the MD, DEA, DAAV, IFGO, ISST, DMSSCT, and IP are associated with solar radio flux with a wavelength of 10.7 cm (f10.7_index), characterizing the solar activity (SA). This means that the Sun is the source of causal relationships, starting with SA and ending with the morbidity of the population on the Earth. But at the same time, the cause of the morbidity can be other bioeffective agents associated with SA, whose contribution to the morbidity can depend on multiple reasons, including the properties of the ionosphere during periods of the polar day and the polar night.

A comparative analysis of the correlations of the same classes of morbidity with geophysical indices, separately for the polar day and for the polar night, showed that there are both general and particular trends in the nature of the relationship between the morbidity and geocosmic agents. There are correlations, which appear only during the polar day: mental disorder (MD), diseases of the arteries, arterioles, and veins (DAAV), the gastritis, the kidney and urinary tract diseases, the complications of pregnancy and the postpartum period, and other diseases. Diseases such as the pneumonia, the ischemic heart disease, and other forms of heart disease without hypertension are correlated with geocosmic agents only during the polar night. There are diseases with a mixed nature of the connections with geocosmic agents during the polar day and the polar night.

Figure 4 shows that during the period of the polar day, dynamics of the monthly values of incidences of the mental disorders, MD, and dynamics of the monthly values of incidences of the diseases of arteries, arterioles, and veins (DAAV) are associated with variations of solar wind parameters such as "sigma-phi-V" and the solar radio emission with wavelength 10.7 cm. Along with these parameters of geocosmic agents, other parameters of IMF and SW, as well as, possibly, their combination and interaction, can make a certain contribution to the modulation of cases of mental disorders (**Table 2**).

The same can be seen in **Table 3**, which shows the links of the diseases of arteries, arterioles, and veins (DAAV) with variations of geocosmic agents, reflecting the complex nature of the effects of physical agents on the diseases of blood vessels.

In general, it can be seen that cases of mental disorders and vascular morbidity are associated with SA, manifested by variability of the solar wind (SW) and IMF during the polar day period. This may mean that, as a result of the interaction of the SW and IMP with the Earth's magnetosphere, physical phenomena generated in the polar cusp region during polar day could contribute to an unstable mental state and vascular disorders.

One can assume that these phenomena have an electromagnetic and wave nature, which determines their bioefficiency. One of the most likely candidates in a wide range of physical phenomena detected in the cusp region is low-frequency pulsations [14].

The pulsations in the spectral range (1–5 mHz) with different morphological properties and, accordingly, with different physical nature are observed at high latitudes ($\Phi > 70°$). It is established that the long-period (T ~ 4–60 min) geomagnetic pulsations observed both in daytime and nighttime hours are typical phenomena on the polar cusp latitudes. The most typical fluctuations of the daytime cusp observed on the earth's surface are specific broadband irregular pulsations of the Pc5 range (f ~ 1.5–5.0 mHz) with an amplitude of the order of 15–60 nT, named by V.A. Troitskaya *ipcl (irregular pulsation cusp latitudes)*. Pulsations of the *ipcl* type are observed almost daily, but their intensity is 3–4 times higher in the summer than in the winter. This fact indicates that the source of *ipcl* pulsations is, in essence, a current generator, which creates the greatest disturbance in the illuminated ionosphere [14].

The daytime geomagnetic pulsations *ipcl* are divided into at least two classes [29]: *np* pulsations having a noise-like character (P = 6–15 min), and relatively regular *vlp* (very long period) pulsations (P = 20–40 min) occurring near the equatorial cusp boundary [29]. In the winter, as a rule, *vlp* pulsations are not detected [14, 29].

In the higher frequency range, the broadband noises from Pc3–4 (10–40 mHz) to ELF choirs (0.3–3.0 kHz) are often observed in the high latitude. The intensity of the Pc3–4 waves in the polar cusp depends on the ionospheric conductivity, which causes a sharp weakening of the waves during the polar night [28, 31, 32].

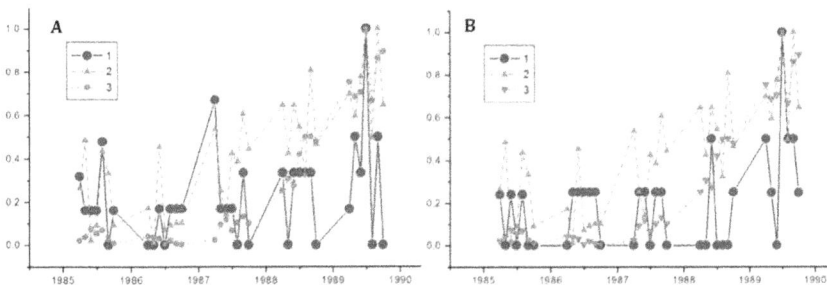

Figure 4.
The relationship between morbidity and geocosmic agents, which appears only in the period of the polar day. A. Dynamics of the monthly values of incidences of the mental disorders, MD (1). B. Dynamics of the monthly values of incidences of the diseases of arteries, arterioles, and veins, DAAV (1); A, B. Dynamics of the monthly average magnitudes of the solar wind parameters "sigma-phi-V" (2) and the solar radio emission with wavelength 10.7 cm (3). X axis: the months of the year from January 1985 to December 1989, where months from March to September are included in the period of the polar day (A, B); Y axis: all normalized parameters.

Period	M ± δ	NM	Pres	Bz	Pr-Den	δ phi	R	f10.7	PC(N)	Makh
PD	1.21 ± 1.20	−0.49	0.43	−0.36	−0.38	0.56	0.41	0.38	0.35	−0.33
PN	1.08 ± 1.08	−0.04	−0.06	0.02	0.07	0.13	0.13	0.16	−0.06	−0.26

NM—count rate of ground based on neutron monitor (counts/s); Pres—atmospheric pressure (mb); Bz-BzGSE—Bz component of interplanetary magnetic field (IMF) in the geocentric solar-ecliptic coordinate systems, nT; Pr-Den—proton density in the solar wind, N/cm^3; δ ph—sigma-phi-V—solar wind angle parameter, deg.; R—sunspot number; f10.7-index of the solar radio flux with wavelength 10.7 cm in solar flux units (s.f.u.), (10^{-22}), Watts/meter sq/hertz; PC(N)—Index of geomagnetic activity in the high latitude; Makh—Magnetosonic mach number = V/ Magnetosonic speed. Coefficient values marked in red color correspond to the level of significance p < 0.05.

Table 2.
Correlation coefficients between monthly values of incidents of the mental disorders (MD) and monthly average magnitudes of the parameters of geocosmic agents during the polar day (PD) and polar night (PN).

Summarizing the descriptions of physical phenomena associated with the processes of the interaction of the solar wind and IMF with the earth's magnetosphere in the polar cusp region, one can see that the polar day differs from the polar night by more diverse geocosmic events. These events are dependent on ionospheric conductivity, which determines diverse phenomena, including amplitude and frequency characteristics of high latitude pulsations.

It has now been established [33–36] that brain rhythms include ultra-slow frequency oscillations (USFO), which are usually not detected by standard electroencephalogram measurements. The frequency range of these oscillations corresponds to very low-frequency pulsations Pc3–4 characteristic of a polar cusp. Among the ultra-slow fluctuations, the rhythm with a period of 15–40 s is remarkable in that the human brain is accompanied by transitions of levels of consciousness, for example, transitions to the hypnotic state. The fluctuations in the decasecond range correspond to the period of fluctuations of the pulsations Pc3, the amplitude and intensity of which are significantly higher during the polar day than in the polar night. It is not excluded that Pc3–4 pulsations can contribute, along with other factors, to the unstable mental state of the residents of arch. Spitsbergen.

Significance (p < 0.05) of correlation coefficients between MD and DAAV (r = 0.40), between MD and DAAV and solar activity (**Tables 2** and **3**) in the polar day and the absence of significance of correlations between these morbidity and SA indices (R, F10.7) during the polar night indicate common causes, which determine the relationship between the morbidity and geophysical agents in the polar day. We assume that such common causes may be geophysical agents associated with the illuminated ionosphere during the polar day. It is possible that geomagnetic pulsations, in the ultralow frequency range, most pronounced during the polar day, could modulate brain and vascular functional activity and, accordingly, certain mental states. In particular, they might suppress the cognitive processing and

Period	M ± δ	NM	δ-By	δ-Bz	Na/Np	δ-phi	δ-theta	R	f10.7
PD	0.70 ± 0.79	−0.52	0.41	0.35	0.38	0.37	0.39	0.51	0.51
PN	0.61 ± 0.81	−0.20	−0.22	−0.06	−0.06	0.04	0.24	0.26	0.31

NM—count rate of ground based on neutron monitor (counts/s); δ-By—sigma By—variability of By-component of IMF, nT; δ-Bz—sigma Bz—variability of Bz-component of IMF, nT; Na/Np—alpha/proton ratio in the solar wind; δ phi, δ-theta—sigma-phi-V, sigma-theta-V—solar wind angle parameters, deg.; R—sunspot number; f10.7-index of the solar radio flux with wavelength 10.7 cm in solar flux units (s.f.u.), (10^{-22}), Watts/meter sq/hertz. Coefficient values marked in red color correspond to the level of significance p < 0.05.

Table 3.
Correlation coefficients between monthly values of incidents of the diseases of arteries, arterioles, and veins (DAAV) and the monthly average magnitudes of the parameters of geocosmic agents during the polar day (PD) and polar night (PN).

promote switching of the brain to its noncognitive "idling" state or activation of default cortical networks whose activity is suppressed during cognitive processing [37, 38].

The different significance of physical agents for different systems of the body can be seen on the basis of the mutually exclusive nature of the connections with similar geocosmic agents in the polar day and in the polar night (**Figure 5, Table 4**).

The only difference in the nature of the connections between these diseases and geocosmic agents is the connection with the Bz-component of IMF. This connection has a negative sign with the incidence of DMSSCT and positive sign with the incidence of ISST in the polar day. Since the negative value of the Bz-component characterizes a high GMA, it can be assumed that GMA, along with other factors, including ultraviolet irradiation, contributes to the incidence of the DMSSCT in the polar day.

On the other hand, excessive irradiation of ultraviolet light during the polar day can inhibit the growth of pathogenic microflora, which causes skin diseases (ISST). But in the polar night, in the absence of ultraviolet light, the growth of pathogenic microflora can increase under the influence of factors associated with the variability of the SW and IMF [39].

The importance of SA for human behavior manifests in the correlations with the cases of injury and poisoning on the way to and from work of the residents of arch. Spitsbergen (**Figure 6, Table 5**). Most likely, this definition hides injuries caused by the state of altered consciousness under the influence of alcohol.

It can be assumed that an increase in SA is accompanied by the neuropsychic arousal, the anxiety, the decrease in health, and the mood, which can be causes provoking the need for alcohol in a certain category of persons. Since the level of SA in the summer and winter periods does not differ significantly, the connection of the frequency of injuries and poisonings on the way to work and from working with SA appears equally on a polar day and on a polar night according to the level of the solar activity.

Monitoring of the daily psycho-emotional state of the healthy volunteers in the settlement Barentsburg (arch. Spitsbergen) during polar day revealed correlations between situational anxiety, mood, activity, and indices of SA of proton fluxes with energy >10 MeV [40, 41]. It was also found that health, the activity, and the mood decreased and the situational anxiety increased under increase of SA and GMA [40, 41]. Thus, one of the causes for the increase in injuries and poisoning could be an arising of the psycho-emotional instability associated with increase in SA.

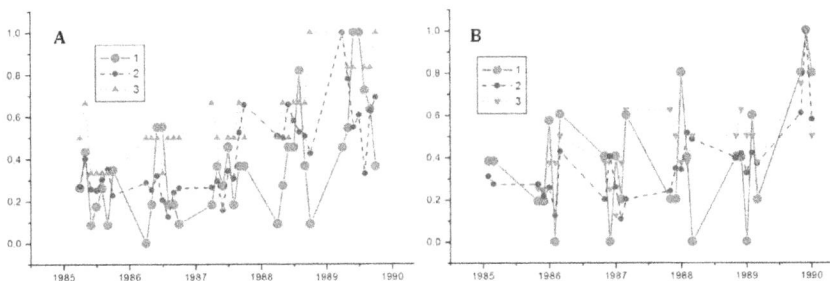

Figure 5.
The mutually exclusive nature of the connections with similar geocosmic agents of the various diseases in the polar day (A) and in the polar night (B). A. Dynamics of the monthly values of incidences of diseases of the musculoskeletal system and connective tissue (DMSSCT) (1), the monthly average magnitudes of interplanetary magnetic field (IMF), nT (2), the monthly average magnitudes of the alpha/proton ratio in the solar wind (Na/Np), (3); B. Dynamics of the monthly values of incidence of the infections of the skin and subcutaneous tissue (ISST) (1), the monthly average magnitudes of IMF (2), the monthly average magnitudes of the Na/Np (3). X axis: the months from January 1985 to December 1989, where months from March to September are included in the spans of the polar day (A); the months from October to February are included in the spans of the polar night (B); Y axis: all normalized parameters.

Period	M ± δ	NM	IMF	FV|<B>|	Bz,GSM	δ-B	δ-By	δ-Bz	Na/Np
Diseases of the musculoskeletal system and connective tissue (DMSSCT)									
PD	3.66 ± 2.48	−0.63	0.44	0.38	−0.39	0.47	0.47	0.44	0.46
PN	3.42 ± 2.95	−0.25	0.26	0.31	−0.21	0.15	0.17	0.21	0.16
The infections of the skin and subcutaneous tissue (ISST)									
PD	1.83 ± 1.61	−0.20	0.23	0.25	−0.12	0.16	0.13	0.12	0.15
PN	1.76 ± 1.27	−0.50	0.57	0.53	0.53	0.50	0.47	0.47	0.50

NM—count rate of ground based on neutron monitor (counts/s); IMF—field magnitude Avg, <F>, nT; FV\|<B>\|—magnitude of average, field vector, \|<B>\|, nT; Bz, GSM-Bz—component INF in the geocentric solar-magnetospheric coordinate systems; δ-B—variability of the magnetic field strength; δ-By—Sigma By—variability of By—component of IMF, nT; δ-Bz—Sigma Bz—variability of Bz—component of IMF, nT; Na/Np—alpha/proton ratio in the solar wind. Coefficient values marked in red color correspond to the level of significance $p < 0.05$.

Table 4.
Correlation coefficients between monthly values of incidents of the diseases of the musculoskeletal system and connective tissue (DMSSCT), the infections of the skin and subcutaneous tissue (ISST), and the monthly average magnitudes of the parameters of geocosmic agents during the polar day (PD) and polar night (PN).

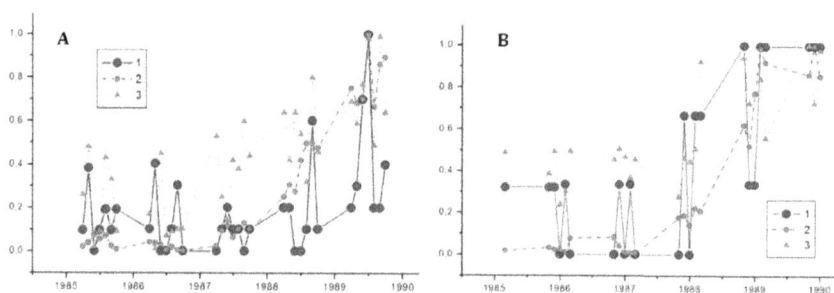

Figure 6.
The stable links between injury rates and poisoning on the way to work and from work with similar geocosmic agents during the polar day (A) and the polar night (B). A, B. Dynamics of the monthly values of incidence of the injury and poisoning on the way to work and from work (1), the monthly average magnitudes of the solar radio emission with wavelength 10.7 cm (2), and the solar wind parameter "sigma-phi-V"(3). X axis: the months from January 1985 to December 1989, where months from March to September are included in the period of the polar day (A); the months from October to February are included in the period of the polar night (B); Y axis: all normalized parameters.

Period	M ± δ	NM	δ phi-	δ theta	Pl beta	AMN	Kp * 10	R	f10.7	PC(N)	MMN
PD	1.75 ± 1.97	−0.52	0.48	0.46	−0.38	−0.43	0.26	0.61	0.58	0.39	−0.43
PN	1.26 ± 1.07	−0.55	0.54	0.51	−0.55	−0.61	0.47	0.67	0.72	0.58	−0.55

*NM—count rate of ground based on neutron monitor (counts/s); δ phi, δ-theta—sigma-phi-V, sigma-theta-V—solar wind angle parameters, deg.; Pl beta—plasma beta (Beta = $[(T * 4.16/10^5) + 5.34] * Np/B^2$; AMN—Alfven mach number (Ma = $(V * Np^{0.5})/20 * B$); Kp * 10-index of geomagnetic activity (GMA); R—sunspot number; f10.7-index of the solar radio flux with wavelength 10.7 cm in solar flux units (s.f.u.), (10^{-22}), Watts/meter sq/hertz; PC(N)—high latitude index of GMA; MMN—magnetosonic mach number = V/Magnetosonic_speed. Coefficient values marked in red color correspond to the level of significance $p < 0.05$.*

Table 5.
Correlation coefficients between monthly values of incidents of the injury and poisoning on the way to work and from work and the monthly average magnitudes of the geocosmic agents during the polar day (PD) and polar night (PN).

The revealed differences in the nature of the links between the morbidity of the population in the Barentsburg during the polar day and the polar night show that the diverse diseases are associated with a combination of separated characteristics of the SV, MMP, GMA, and SA, the significance of which for the morbidity varies with the season.

4. Conclusion

A geophysical feature of the arch. Spitsbergen is its location in the cusp region— a kind of funnel on the dayside of the magnetosphere with near zero magnetic field magnitude. The open field lines of the cusp are connected with those of the inter-planetary magnetic field (IMF), which allows the shocked solar wind plasma of the magnetosheath to enter the magnetosphere and to penetrate the ionosphere.

In the cusp areas, the impacts of the solar wind (SW) on the Earth's mag-netosphere manifest most strongly, and multiple phenomena originating as consequences of such interactions are referred to as the space weather. The magne-tosphere and the ionosphere of the Earth are sources of electromagnetic oscillations and waves, many of which are detected in the form of radiation outside the region of generation, in particular, on the surface of the Earth.

The feature of the cusp is the existence of the geomagnetic pulsations not only in the period of geomagnetic disturbances but also during the quiet period. One can see that narrow band waves at frequencies 0.2 to 3 Hz are a permanent feature in the vicinity of the polar cusp. The waves have been found in the magnetosphere adjacent to the cusp (both poleward and equatorward of the cusp) and in the cusp itself. It is an established fact that the daytime polar cusp latitudes are typically characterized by long-period (T ~4–60 min) geomagnetic pulsations observed both in daytime and nighttime hours. Diverse radiation (from ULF to VHF) and waves in the field of polar cusp, covering the entire range of the body rhythms, give credit for studying the effects of space weather in the field of polar cusp. The study of the dependence cases of diseases on effects of space weather has shown that diverse forms of morbidity varied synchronously and they are associated with variations of space weather agents. Assessment of the relationship between the dynamics of the monthly morbidity in Russian settlements and indicators of space weather revealed that, practically, all forms of morbidity are associated with solar activity: with F10.7 index, with variations of solar wind parameters, and with indices characterizing the local geomagnetic activity in the polar cusp.

It has been found that mental disorders are associated with the variability of the solar wind and the radio emission of the Sun, as well as the frequency of injuries and poisoning at the work and at the home. A high degree of association of the diseases of arteries, arterioles, and veins with the parameters of the solar wind and the geomagnetic indices, characterizing the local geomagnetic activity in the polar cusp, was shown.

A high sensitivity of the female organism to variations of space weather in the polar cusp was revealed. This phenomenon is manifested in the increase of preg-nancy complications, cases of inflammation of the genital organs, etc., according to the increase in geomagnetic activity in the polar cusp.

The revealed differences in the nature of the links between the morbidity of the population in the Barentsburg during the polar day and the polar night show that the diverse diseases are associated with a combination of separated characteristics of the SV, MMP, GMA, and SA, the significance of which for the morbidity varies with the season.

However, it has been found that certain diseases are associated only with the polar day or with the polar night. This allows selecting the physical agents that could modulate morbidity rate in the alternative season. In particular, agents such as long-period oscillations, with the frequency range that coincides with the range of the ultraslow fluctuations of the constant potential (USFCP) in the brain, could modulate the morbidity of the MD and DAAV in the polar day.

The absence of solar radiation during the polar night, such as UV radiation, and the association of the incidence of the inflectional diseases of skin with GMA only during the polar night indicate the role of UV in suppressing the growth of pathogenic microflora. Correlations of the inflectional diseases of skin with GMA in the absence of UV radiation demonstrate the significance of GMA for the microorganism growth.

In general, it should be noted that, probably, many of the bioeffective agents associated with CA were left out of consideration. The health of the population most likely depends on a combination of geophysical agents, some of which are simply not registered and are not reflected in the indicators of the database (OMNI). On the other hand, the state of the human body during the periods of the polar day and the polar night may also differ in sensitivity to the effects of geophysical agents. In general, the polar day is characterized by a larger number of influencing physical agents on the human body, than the polar night.

The found links between the morbidity of the population and the peculiarities of space weather will make it possible to develop prognoses of the morbidity for preventive measures aimed at reducing the morbidity in high latitudes.

The task of studying the labor activity in the difficult arctic conditions demands the need to develop criteria for determining the mental state of a person and his working capacity, as well as predicting a shift in the functional state of the CNS. The solution of such a problem should take into account the possibility of modulation of the mental and of the physiological state of people of the dangerous professions by the high latitude geocosmic agents, the effects of which might also express in the seasonal manifestation of morbidity.

Author details

Natalia K. Belisheva
Research Centre for Human Adaptation in the Arctic, Branch of the Federal
Research Centre "Kola Science Centre of the Russian Academy of Science",
International Academy of Ecology, Man and Nature Protection Sciences,
Apatity, Murmansk Region, Russia

*Address all correspondence to: natalybelisheva@mail.ru

IntechOpen

References

[1] Tsyganenko NA, Russell CT. Magnetic signatures of the distant polar cusps: Observations by polar and quantitative modelling. Journal of Geophysical Research. 1999;**104**:24939-24955

[2] Mitch Battros. Astronomers Have Discovered how Earth's Magnetic Field Survives Intense Solar Storms. 2018. Available from: https://scienceofcycles. com/astronomers-have-discovered-how-earths-magnetic-field-survives-intense-solar-storms/

[3] Available from: http://pluto.space. swri.edu/IMAGE/glossary/cusp.html

[4] Solar System and Beyond. Earth's Magnetosphere. March 21, 2011. In: Zell H, editor. Credit: NASA/ Goddard/ Aaron Kaase. 2017. Available from: https://www.nasa.gov/mission_pages/ sunearth/multimedia/magnetosphere. html

[5] Cowley SWH. Magnetosphere-ionosphere interactions': A tutorial review. In: Ohtani S-I, Fujii R, Hesse M, Lysak RL. editors. Magnetospheric Current Systems. Geophysical Monograph Series. Vol. 118. Washington, DC: American Geophysical Union. 2000. pp. 91-108. Available from: https://doi.org/10.1029/GM118p0091 (First published: 01 January 2000)

[6] Holtet JA, Egeland A, editors. In: The Polar Cusp. Nato Science Series C. Series Volume 145. Netherlands: Springer; 1985. 436 p. DOI: 10.1007/978-94-009-5295-9

[7] Natural Resources Canada. What Is Space Weather? [Internet]. 2017. Available from: http://www. spaceweather.gc.ca/sbg-en.php#gen-1

[8] Crosby NB, Rycroft MJ, Tulunay Y. Overview of a graduate course delivered in Turkey, emphasizing solar-terrestrial physics and space weather. Survey Geophys. 2006;**27**:319-364

[9] Kintner PM, Jr. A Beginner's Guide to Space Weather and GPS. Cornell University Lecture Notes; 2008. 12 pp. Available from: https://gps.ece.cornell. edu/SpaceWeatherIntro_ed2_10-31-06_ed.pdf

[10] Klimenko VV. VHF radio emission of the polar ionosphere. Specialty 25.00.29—"Physics of the atmosphere and hydrosphere" [thesis for the degree of candidate of physical and mathematical sciences]. Irkutsk; 2002. 129 p

[11] Sato Y, Ono T, Kumamoto A, Sato N, Ogawa Y, Kadokura A, et al. Ground-based observation of MF auroral radio emissions in the polar cap and cusp regions. 2008. Available from: http:// stpp1.geophys.tohoku.ac.jp/; http:// www.sgepss.org/sgepss/sookai/124/ html/program/pdf/B006/B006-17.pdf

[12] LaBelle J, Hughes JM. Observations of auroral roar emissions at polar cap latitudes' results from the early polar cap observatory. RadioScience. 2001;**36**(6):1859-1868. DOI: 10.1029/1999RS002309

[13] Gorbachev OA, Truhan AA. Ion-acoustic turbulence of an ionosphere as a source VLF radio emissions of type auroral hissings. Scientific Bulletin of the Moscow State Technical University of Civil Aviation. 2005;**93**:120-126

[14] Kozyreva OV. Wave structure of magnetic storms. [thesis, dissertations for the degree of Doctor of Physical and Mathematical Sciences]. Moscow; 2013

[15] Belisheva NK, Vinogradov AN, Vashenyuk EV, Tsymbalyuk NI, Chernous SA. Biomedical research on Svalbard as an effective approach to

studying the bioefficiency of space weather. Herald of the KSC RAS. 2010;**1**:26-33

[16] Belisheva NK, Popov AN, Petukhova NV, Pavlova LP, Osipov KS, Tkachenko SE, et al. Qualitative and quantitative characteristics of geomagnetic field variations with reference to functional state of human brain. Biophysica. 1995;**40**:1005-1012

[17] Soroko SI, Bekshaev SS, Belisheva NK, Pryanichnikov SV. Amplitude-frequency and spatio-temporal reorganizations of the bioelectric activity of the human brain with strong disturbances of geomagnetic activity. Vestnik of the Far East Branch of the Russian Academy of Sciences. FEB RAS Publisher — Central Scientific Library. FEB RAS. 2013;**4**:111-122

[18] Chernouss S, Vinogradov A, Vlassova E. Geophysical hazard for human health in the circumpolar auroral belt: Evidence of a relationship between heart rate variation and electromagnetic disturbances. Natural Hazards. 2001;**23**:121-135

[19] Rozhkov VP, Belisheva NK, Martynova AA, Soroko SI. Psycho-physiological and cardiohemodynamic effects of solar, geomagnetic, and meteorological factors in humans under the conditions of the Arctic region. Human Physiology. 2014;**40**(4):397-409

[20] Belisheva NK, Konradov SA. The value of geomagnetic field variations for the functional state of the human body in high latitudes. Geophysical Processes and Biosphere. 2005;**4**(1/2):44-52

[21] Belisheva NK, Konradov AA, Janvareva IN. Impact of the high latitude geomagnetic field variations on the human cardiovascular system. In: Atkov OY, Gurfinkel YI, editors. Proceeding of an International Scientific Workshop "Space Weather Effects on biological

System and Human Health held in Moscow", Russia, February 17-18, 2005. 2006. pp. 86-87. ReproCENTR M, Moscow

[22] Belisheva NK, Megorsky VV. The role of variations in high-latitude geophysical agents in the dynamics of the prevalence of socially significant diseases in the Arctic. In: Gorbaneva SA, Frolova NM, editors. Problems of Preserving Health and Ensuring the Sanitary and Epidemiological Well-being of the Population in the Arctic: Proceedings of the Scientific and Practical Conference with International Participation. SPb: LLC IPK Costa. 2017. pp. 37-43

[23] Zaitsev AN. Spatio-temporal characteristics of polar geomagnetic disturbances [thesis for the doctor of physical and mathematical sciences]. Code of specialty HAC: 04.00.23. Moscow. 2000. 323 p

[24] Gwal AK, Bhawre P, Mansoori AA, Khan PA. Study of GPS derived Total Electron content and scintillation index variations over Indian Arctic and Antarctic Stations. Journal of Scientific Research. 2013;**5**(2):255-264. DOI: 10.3329/jsr.v5i2.12724

[25] McPherron RL. Magnetic pulsations: Their sources and relation to solar wind and geomagnetic activity. Surveys in Geophysics. 2005;**26**(5):545-592

[26] Troitskaya VA, Gul'elmi AV. Geomagnetic micropulsations and diagnostics of the magnetosphere. Space Science Reviews. 1967;**7**(5-6): 689-768

[27] Kato Y, Saito T. Morphological study of geomagnetic pulsations. Journal of the Physical Society of Japan (Suppl. A)— Part II. 1962;**17**:34-39

[28] Chugunova OM. Geomagnetic Pc3-4 pulsations in the polar cap [thesis]. In: Schmidt OY, editor. Institute of Physics of the Earth. RAS Moscow. 2006. 108 p

[29] Kleimenova NG, Nikiforova NN, Kozyreva OV, Michnovsky S. Longperiod geomagnetic pulsations and fluctuations of the atmospheric electric field intensity at the polar cusp latitudes. Geomagnetism and Aeronomy. 1996;**35**:469-477

[30] Spitsbergen/Svalbard. Polar Night, Polar Day [Internet]. 2014. Available form: https://www.spitsbergen-svalbard.com/2014/02/19/polar-night-polar-day.html

[31] Chugimova OM, Pilipenko VA, Engebretson MJ, Fukunishi HP3. Pulsations deep in the polar cap: A study using Antarctic search-coil magnetometers. In: "Problems of Geocosmos", Proc. of 4-th International Conference, St-Petersburg. 2002. pp. 111-115

[32] Chugunova OM, Pilipenko VA, Engebretson MJ, Rodger A. Pc3-4 pulsations in the polar cap Proc. of the 26th Annual Seminar "Physics of Auroral Phenomena", Apatity, 2003. p. 33

[33] Vanhatalo S, Voipio J, Kaila K. Full-band EEG (fbEEG): A new standard for clinical electroencephalography. Clinical EEG and Neuroscience. 2005;**36**(4):311-317. DOI: 10.1177/155005940503600411

[34] Aladzhalova NA. Psychophysiological aspects of a super-slow rhythmic brain activity. Moscow: Science. 1979. 214 p

[35] Ilyukhina VA. Analysis of brain neurodynamics in different ranges of the amplitude-time spectrum of bioelectric activity. Human Physiology. 1979;**5**(3):467-499

[36] Shvets-Teneta-Gury TE. Bioelectrochemical activity of the brain. Moscow: Science. 1980. 208 p

[37] Pfurtscheller G, Schwerdtfeger A, Brunner C, Aigne C, Fink D, Brito J, et al. Distinction between neural and vascular BOLD oscillations and intertwined heart rate oscillations at 0.1 Hz in the resting state and during movement. PLoS One. 2017:1-13. DOI: 10.1371/journal.pone.0168097

[38] Steyn-Ross ML, Steyn-Ross DA, Sleigh JW, Wilson MT. A mechanism for ultra-slow oscillations in the cortical default network. Bulletin of Mathematical Biology;**73**(2):398-416. DOI: 10.1007/s11538-010-9565-9

[39] Zavadskaya TS, Mikhailov RE, Belisheva NK. Analysis of the contributions of geophysical agents and endogenous microflora in the incidence of men with diseases of the genitourinary system in the Kola North. Journal of Ural Medical Academic Science. 2018;**15**(2):162-175. DOI: 10.22138/2500-0918-2018-15-2-162-175

[40] Belisheva NK, Pryanichnikov SV, Solovyovskaya NL, Megorsky VV. Arch. Spitsbergen—Polygon for Analog Studies of the Effects of Cosmophysical Agents on the Human Body. Russia: Herald of the KSC RAS; 2017;**4**:21-28

[41] Belisheva NK, Martynova AA, Pryanichnikov SV, Solov'evskaya NL, Zavadskaya TS, Megorsky VV. Connection of the Parameters of the Interplanetary Magnetic Field and the Solar Wind in the Polar Cusp Region with the Psychophysiological State of the Inhabitants of Arch. Spitsbergen. Herald of the KSC RAS; 2018;**4**:5-24. DOI: 10.25702/KSC.2307-5228.2018.10.4.5-24

Chapter 6

Indigenous Communities in the Arctic Change in Socio-Economic and Environmental Perspective

Violetta Gassiy

Abstract

In recent decades, the world has undergone significant changes in the environment, which have led not only to economic losses but also to a deterioration in the quality of human life, a change in the usual way of life. The Arctic today is in the focus of geopolitical and economic interests, the impact on the region of global warming. The ice retreats giving humanity new transport corridors, thereby attracting new participants from non-Arctic countries. Japan, China, and South Korea are interested in developing the Northern Sea Route for the delivery of goods and the development of economic ties between Europe and Asia. However, the importance of this route is also connected with transportation of hydrocarbons and other minerals extracted in the Arctic. Industrial development is a priority for the Arctic countries, and climate change makes remote areas of the subsoil more accessible. Especially this issue should be considered for Russia, where the development of the Arctic is experiencing a third wave and this process affects the interests of state, business, and population including indigenous communities, whose number is more twice than in the rest of the world (2.8 million residents in the Russian Arctic with approximately 4 million people in the Arctic totally).

Keywords: indigenous community, Arctic, climate change, territories of traditional nature use, sustainable development, Russia, Yakutia

1. Introduction

At the beginning of this century, the topic of global climate change became of particular relevance for the regions of the Arctic and the North. This problem is actual in modern conditions. The Arctic climate changes faster than any other part of the world; this is the only highly integrated system in this belt; changes in the Arctic will have a big impact on other parts of the world. The Arctic will become an increasing center of world attention. Over the past few decades, the average annual temperature due to an increase in the average winter temperature in the Arctic has grown two times faster than elsewhere, causing the melting of sea ice and permafrost and a reduction in the snow period. The consequences of global warming in the Arctic are already obvious and numerous. Modern climate changes significantly affect coastal communities, species diversity of animals and plants, human health and welfare, as well as the economy and infrastructure of the Arctic regions. Global warming is the process of gradual growth of the average annual temperature of the surface layer of the Earth's atmosphere and the World Ocean, due to all sorts of reasons (increase in the concentration of greenhouse gases in the Earth's atmosphere, changes in solar or volcanic activity, etc.). Global warming will change the habitats of many species of terrestrial and marine flora and fauna. The most large-scale changes will be felt by the indigenous peoples of the North, whose life is inseparably linked with the natural environment. As the permafrost is thawing, the threat of destruction of buildings, roads, pipelines, airports, and other infrastructure increases, which in a number of cases will lead to significant economic losses, deterioration in the quality of drinking water supply, social tension, forced migration, and, as a result, an increase in the number of infectious and noninfectious diseases, including mental disorders and psychosomatic and addiction diseases. Indigenous peoples of the North are the most vulnerable category of the population to the climate negative impact in the Arctic. Limiting the possibility of using bioresources as a result of hunting and reindeer herding, fishing, and gathering, as well as reducing the safety of movement when the parameters of ice and weather conditions change significantly, increases the risks to health and life and, possibly, in the future, threatens the very existence of some nationalities and cultures.

Gradually, in countries the understanding comes that the nature is the original environment of human life, but not capital, which should be used in economic circulation. Preservation of this environment is becoming one of the main tasks of state policy based on the principles of energy efficiency and resource saving. For example, in the Russian Federation, such basic documents as the Strategy of Ecological Safety of Russia [1], the state program on energy efficiency and development of energy [2], etc. were adopted. However, despite the billions of dollars invested by developed countries in greening the economy, the development of innovative technologies, and the reduction of greenhouse gases, there are still no visible effects on a global scale, and in fact the world is facing a degradation of the natural environment. As Nobel laureate academician Vladimir Kotlyakov notes, our planet is experiencing an era of global warming. The increase in global air temperature in the last century was slightly more than 0.7°C. However, over the past 30 years, this growth has increased, which is especially reflected over the continental regions of Eurasia and North America and most of all in the Arctic [3]. The current model of the functioning of the world economy allows us to make disappointing forecasts: the growing population of the Earth will be able to supply the products of consumption only with the increase of production, the improvement of technologies, and, unfortunately, the destruction of the biosphere.

Figure 1 demonstrates the anomalies of temperature values in the Northern Hemisphere, including the Arctic. This gives grounds to predict the increasing influence of negative factors on the ecosystem of this region, as well as on the life of the indigenous population. Certainly, climate change is a particularly important issue in the context of the development of the Arctic and the indigenous communities that inhabit it. Indigenous peoples also have their own observations related to climate change, since no one can see better what is happening now in the North, and there are significant shifts in their strategies for adapting to these changes. Traditional knowledge is a valuable resource that can and should be used in various fields of exploration and development of the Arctic. Unfortunately the representation of indigenous peoples in international governance structures does not guarantee that traditional knowledge is entirely engaged in evidence-based policy making and that traditional knowledge is not always valued as an equal source of knowledge by some relevant scientific bodies [4]. Hundreds of years of tribal communities' observations over the changes in the Arctic, the formation of ideas about the laws of nature, beliefs in the "living land of ancestors" give today the opportunity to transform traditional knowledge into the daily practice of government, business,

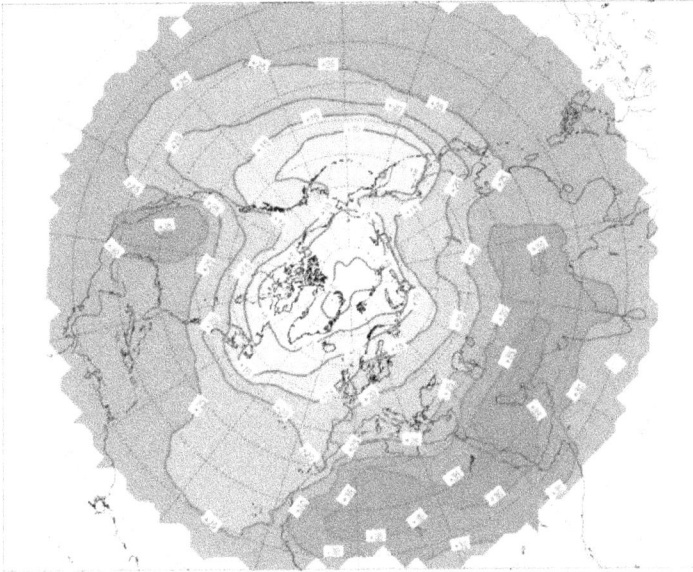

Figure 1.
Map of monthly values and anomalies of meteorological values in the northern hemisphere for June 2018 (source: https://meteoinfo.ru/anomalii-tabl3).

and scientists in the extreme North and integrate it with modern technologies. According to the Paris Climate Change Agreement, indigenous peoples and local communities are recognized as the important actors in building a world that is resilient in the face of climate impacts [5].

2. Factors, risks, and challenges to the indigenous communities in climate change in the Arctic

We can rightly call the Arctic zone a "locomotive" of the modernization of the Russian economy [6]. In this vein, state policy is being drawn up, investments are attracted, and projects are being implemented to extract natural resources (gas, oil, gold, rare earth metals, etc.). Almost every one of these projects implemented in the northern regions of the country, one way or another, affects the territories of traditional nature use—the habitat of indigenous peoples of the North. Therefore, the issue of research and assessment of changes in these territories under the impact of climate change and industrial development is very relevant, since it has a multi-factorial specificity, centered on the unique culture of the northern people, its traditions, and its customs. In Russia, indigenous peoples of the North, as a rule, live in the rural areas of the Arctic zone, which population, according to the Federal State Statistics Service, declines annually. Therefore, it is important to study the changes in these territories and develop policies aimed at preserving not only local communities as a carrier of culture and traditions of northern peoples but also traditional economic activities (reindeer herding, fishing, hunting, etc.), since the reindeer herding is the basis of the traditional culture of the North (**Figure 2**).

The future of the Arctic territories is connected, on the one hand, with the expansion of the zone of industrial development and the extraction in deposits and on the other hand the increasing pressure on the unique ecosystem of the Arctic, the changes in the territories of traditional nature use, the transformation of indigenous

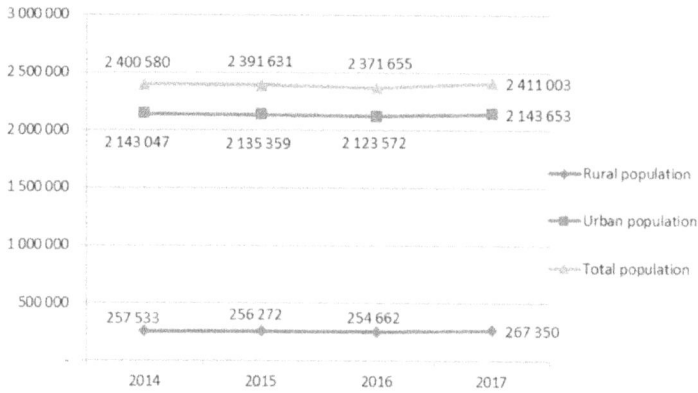

Figure 2.
The numbers of the permanent population of the land territories of the Arctic zone of the Russian Federation as of January 1, 2018 (number of people) [7].

Education organizations, units	1735
Number of medical treatment and prophylactic organizations, units	2045
Number of cultural and leisure type organizations, units	834
Libraries and museums, units	542
Number of sports facilities, units	10,161
Hospitality facilities and accommodation	1123
Shops and supermarket, units	28,364
Restaurants and cafes	3773
Settlements with post office, units	1735
Commissioning of residential buildings, square meters	1,496,550
Number of people living in dilapidated houses	192,411
Extension of a street water supply network, meters	7,566,841.0
Including in need of replacement, meters	2,411,098.0
Number of enterprises for utilization and neutralization of domestic and industrial waste, units	151

Table 1.
Indicators of the social development level of territories of traditional nature use in Siberia and the Far East of the Russian Federation in 2017 [9].

population's way of life, and tribal communities under the influence, including climate change. In **Table 1** the main indicators describing the territories of traditional residence and traditional economic activity of the indigenous peoples of Russia are presented. This type of territory is located in 21 regions of the Russian Federation with reindeer pastures, hunting grounds and rich fishing opportunities, and gathering of wild plants on a total area of 994.2 million hectares, including lands used directly as reindeer pastures—407.0 million hectares [8].

Climate change leads to the transformation of the traditional way of life and also forces regional and local governments to seek new approaches to managing these changes, allowing them to adapt and adequately respond to emerging challenges. Prospects for the revitalization of the industrial development of the North in the future involve the withdrawal of an increasing number of lands of traditional nature use for inclusion in economic circulation. Undoubtedly, industrial development of indigenous peoples' habitats at the present time determines the prospects for their further socioeconomic and ethno-cultural development. Considering the strategic nature of the state interests in the Arctic region and the attention paid to the development of deposits in Siberia and the Far East, it is necessary to devote harmonization of interests of industrial development of these territories and preservation of the habitat of indigenous communities, creation of mechanisms for interaction of task forces on optimization of economic, and social and environmental interests of all stakeholders in the territories of traditional nature use [10].

2.1 Climate change impacts on the traditional way of life

The impact of climate change on indigenous peoples is diverse. This is especially reflected in health and the traditional way of life. Health as a factor in the well-being of indigenous peoples worsens, which shows itself in a high level of mortality with relatively high birth rates, problems with alcoholism, and diseases of the digestive system due to poor-quality drinking water. Significant climate change resulting in the increase of natural disasters, abnormal winter and summer temperatures, floods, mudflows, and landslides increases the number of deaths from unnatural causes, injuries, and subsequent health problems. Climatic changes are also the cause of more serious phenomena, as the deterioration of the parasitic and epidemiological situation. Degradation of permafrost in areas where this type of soil has been preserved for centuries, and on the basis of which the habitat of indigenous peoples and their feeding systems has been formed, leads to catastrophic consequences. Moreover, changes in the permafrost sometimes have unexplained causes, which raise an active discussion in the scientific community. So, in 2014 in Yamal, a giant dip of a soil of unknown origin was discovered. A huge funnel was noticed by helicopter pilots who serve the oil and gas fields on the Yamal Peninsula. The fault is located next to the Bovanenkovo gas field (Yamal LNG), one of the largest in Yamal—the place of one of the most innovative projects of modern Russia for liquefied gas production jointly implemented with Italy, France, Japan, and China (**Figure 3**).

Later Russian scientists from Yamal managed to descend for the first time to the bottom of this dip—to a depth of 200 m. The hole has a cone-shaped view with dimensions of 60 and 40 m. They took more than a dozen samples for chemical analyses, including ice and soil. It turned out that the Yamal "black hole" from the

Figure 3.
Yamal hole in 30 km from the Bovanenkovo gas field. (source: https://www.moya-planeta.ru/news/view/uchenye_vpervye_issledovali_dno_yamalskoj_voronki_8251/)

inside is covered with a layer of ice of unknown composition, which has yet to be investigated in the laboratory. Analyses of air inside the funnel revealed the absence of harmful impurities and dangerous gases—on the basis of this fact, scientists concluded that in a mysterious earthly failure, a new life could arise in time. The scientists noted that they failed to solve the main riddle—how the process of a mysterious holes' formation was going on in the Yamal land. The most authoritative experts consider these holes to be the result of the process of degassing the permafrost due to global warming [11]. Currently, the problem of tundra transformation under the climate change factors is becoming a significant threat to the traditional forms of economic activity, especially reindeer herding. The formation of thermokarst lakes, the degradation of biota, and the waterlogging of significant areas of the tundra during the summer period are risk factors and cause deer to change routes, and in the spring and autumn, a phenomenon such as ice, which is ruinous for reindeer herding, began to increase. Due to the steady increase in the amount of precipitation in recent years, a deeper snow cover is formed, creating difficulties for animals to hoof the reindeer moss. At the same time, the late arrival of colds led to difficulties in the transition of reindeers to winter pastures (**Figure 4**).

In the northern regions of Russia in recent years, there have been no isolated cases in which thousands of reindeers perished from hunger. The increase in the mean annual temperature is detrimental to the regions of permafrost, where the centers of anthrax are revealed during thawing. In the summer of 2016 on the territory of Yamal, an outbreak of anthrax was caused by an abnormal heat. The most dangerous infection was safely suspended in the permafrost for 75 years. The most objective cause of the outbreak was called climate warming. Abnormal heat in the tundra to +35°C kept for more than a month. Comprehensive measures were taken to protect Yamal reindeer herders from dangerous diseases. All the livestock of the deer are vaccinated; the animals are fitted with chips. Vaccination is conducted among the tundra population and specialists from the risk group: in 2017, about 8.2 thousand people were vaccinated in the region, and the entire number of reindeer and more than 730 thousand animals were vaccinated against

Figure 4.
Thermokarst lake in tundra with landscape degradation near Vorkuta, Komi Republic, Russia (photo: V. Gassiy).

anthrax. Forty-two thousand representatives of the indigenous peoples of the North—14,000 of them live in a traditional nomadic way of life—and the largest reindeer herd in the world live in the territory of Yamal, so the ecological component plays an important role in preserving the traditional economic activities of indigenous peoples [12]. One of the main threats is the change in the water regime of rivers. Most of the modern settlements in the North are located on the banks of rivers. In recent years, spring floods have sharply increased, floods have become more frequent, and the processes of erosion of shores have accelerated, which bring great disasters to the population. For example, in the regions of northern Yakutia, the banks of the rivers Lena, Yana, and Anabar collapse under the influence of high temperatures and melting of permafrost, which leads to shallowing of rivers, a change in the relief of the bottom. As a consequence of these processes, boats of local fishermen cannot sail along the riverbed; the fish does not go far downstream; thus indigenous peoples are deprived of the type of product that forms the basis of their food ration. Reduction of fishing takes place together with a decrease in the level of production of hunting objects (wild reindeer changes migration routes; the number of fur-bearing animals decreases; because of warming, the meat of a wild animal is often affected by a viral infection or parasites), i.e., we are talking about the problem of access to traditional types of resources.

Access to resources is closely linked to security, which is provided by traditional knowledge, accumulated for millennia. But the transformations that are taking place change the reality; the representatives of indigenous peoples are increasing in situations where their practice, experience, and knowledge cannot help them. This leads to an increase in the number of accidents, especially those associated with late freeze-up, ice, and early floods. One of the consequences is the restriction of access to traditional food. In addition to the above factors, one of the reasons is the deterioration of storage conditions. In recent years, the quality of food has sharply deteriorated. So, in the Bulunsky District of Yakutia, local residents often face the problem of phimosis (*cysticercosis*) caught from the Lena River. It should be noted that a similar problem is a characteristic of other regions of the Arctic where indigenous peoples eat fresh or slightly salted fish [13]. In 2016, Federal Service for Veterinary and Phytosanitary Surveillance in the Komi Republic during federal monitoring sampled liver and kidney samples of slaughtered animals belonging to the reindeer herding enterprises of Intinsky and Usinsky districts. Sixty-four samples were examined, of which 52 results were found with excess of mercury—the maximum permissible level was exceeded by 0.9 mg/kg—and 43 results with excess of cadmium, the maximum permissible norm is exceeded by 8.3 mg/kg [14]. In addition, in the liver samples, an excess of the normative indices of dioxins was detected—the maximum permissible rate was exceeded by 8.3 times. However, meat and other offal (with the exception of kidneys and liver) do not contain dangerous chemical pollutants and do not pose a danger to citizens. Accumulation of toxicants in the liver and kidneys of animals is due to the physiological properties of these organs, which are biological filters of organisms. Dioxins are formed in a number of industrial and natural processes, for example, in the production of chlorine and pesticides, burning fuel and debris, and forest fires. Cadmium and mercury pollute the environment both for natural and as a result of industrial activities. In particular, heavy metals pollute the environment during the smelting of nonferrous metals and other processes in the mining industry. It is believed that the northern communities of plants and animals tend to accumulate persistent contaminants, as they have a number of properties necessary for this, including the characteristics of the climate (preventing the destruction of substances) and food chains that are distinguished by a small variety of plant and animal species. According to the world scientific data, some traditional food of the inhabitants of the northern regions of the planet (Alaska, Greenland, Scandinavian Peninsula, Far North of Russia) have a high content of harmful chemicals. Such types of food include meat and fat of marine mammals, reindeer offal, and others [15]. In this way, there are more and more people who are forced to refuse from the consumption of raw fish, which often turns out to be infected with phimosis and other diseases. As a result, the probability of losing certain cultural traditions is growing, since food is an integral part of the traditional way of life and culture.

It is also necessary to say about the impact of climate change on the health of indigenous peoples. In recent years there has been an increase in mortality in the Arctic. Almost every year there are floods, with every third year—with disastrous consequences and deaths. The number of hits to hospitals increased due to sunstroke, dehydration, pressure drop, etc. Surface water pollution increased, both from floods and melting of permafrost. This leads to an increase in intestinal diseases, especially in the period of floods. Also, in Arctic regions, there is increasing cases of oncological diseases [16]. Some experts attribute this to a more intensive chlorination due to the deterioration of water quality. The prolonged exposure to increased concentrations of chlorine and its constituents, according to doctors, increases the risk of cancer. Warming has widened the areas of spread of diseases, the carriers of which are insects or mites that spread to all new territories. One of the main risk groups for climate change is the children. In northern regions, up to 70% of children have deviations in health status. The incidence of children in the northern regions is significantly higher than the national average. Over the past 10 years, they tend to grow. Children of the North and children of other regions are in unequal starting conditions of life. Under the influence of unfavorable climatic factors and polluted environment, the age development of the immune system falls behind in children of the North for 2–5 years. Thus, for indigenous peoples of the Arctic, the warming of the climate and the associated lengthening of the season, during which the sea is not covered by ice, a decrease in the surface and thickness of sea ice, changes in the migration routes of wild reindeer and their food base, and a drop in the number of marine animals may lead to a reduction in traditional craft. This, in turn, will lead to a violation of traditional food. The indigenous inhabitants of Alaska and Greenland, Chukotka, and Yamal are already recording the negative effects of climate warming, which appeared in a decrease in thickness and an earlier opening of sea ice. These circumstances make it more difficult to hunt and lead to an increase in the number of injuries, which is already the cause of a significant number of deaths among indigenous peoples of the North [17].

Figure 5 shows the riverbed of the Anabar River near the village of Saskylakh in the northwestern part of Yakutia. Fishermen are forced to manually drag the boat a few kilometers downstream to reach the fairway (**Figure 6**).

2.2 In a dialog with indigenous communities for better understanding of climate change in the Arctic: the Yakutia case

In 2017, an expedition aimed to the research on socioeconomic and environmental problems of the Arctic indigenous communities was organized by the financial aid of the Russian Fund for Basic Research (RFBR) to the Anabar National (Dolgan-Evenk) ulus (district) and Ust-Yanskiy region in Yakutia. These areas belonged to the compact residents of the indigenous peoples of the North. The study allowed to determine the attitude of the local population to traditional activities and to identify the socioeconomic problems of the territories and environmental threats to indigenous communities in the context of climate change. In the

Figure 5.
Shallowing of the Anabar River in Yakutia (photo: V. Gassiy).

Figure 6.
Collapse of the riverbank of Yana due to permafrost melting, Yakutia (photo: V. Gassiy).

structure of the respondents in the Anabar area, representatives of indigenous peoples were Evenks 43 people (33%) and Dolgans 71 people (55%) (**Figure** 7) (**Table** 2).

It is worth noting that this ratio between men and women, when the number of women prevails, is typical for indigenous communities, since it is associated with the high mortality of men engaged in traditional crafts: hunting, fishing, and reindeer herding. In addition, we can add problems of alcoholism reducing life expectancy, as well as chronic diseases caused by the harsh climate. As a result of the survey, residents of indigenous communities noted the following socioeconomic problems in their places of residence:

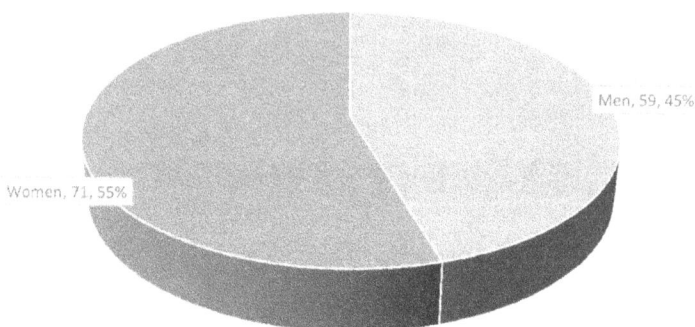

Figure 7.
Ratio of men and women in the total number of respondents.

Indigenous community	Respondents	Share of respondents from the total number, %
Yuryung-Khaya	29	22
Saskylakh	101	78
Total	130	1000

Table 2.
Distribution of respondents who participated in the survey, by settlements in the district.

1. High prices for food products, 22.5%

2. The lack of jobs, 20.2%

3. Low level of income, 19.7%

4. Old state of housing and communal services, 19.1%

5. Poor transport accessibility, 9.0%

6. Low level of medical services, 6.5%

7. Low level of equipping educational institutions, 3.1%

As can be seen from the survey results presented, the majority of the respondents connect the socioeconomic problems of the territory with the lack of a stable income, the need for employment, and the underdeveloped infrastructure. In this regard, the implementation of investment projects for the industrial development of territories can create additional jobs for the local population. It should be noted that in the experience of some Russian regions, there are examples of the implementation of a targeted policy for the local labor market development. For example, for several years in the Republic of Sakha (Yakutia), JSC Almazy Anabara (Alrosa group) has been implementing the educational program, which provides training for the company's interests and the residents of indigenous communities, where an investment project on the extraction of minerals starts. The survey made it possible to determine the list of sociocultural problems that concern the local population:

1. Increase in morbidity and mortality of the population, 20.7%

2. Loss of communication between people and their culture, traditions, 18.3%

3. Alcoholism, 18.3%

4. The lack of organized forms of leisure, 14.1%

5. Problems of selling traditional craft products, 12.3%

6. Outflow of youth, 12.0%

7. Crime rate, 4.5%

It should be noted that the majority of respondents attributed an increase in morbidity and mortality of the population with active industrial development of territories of traditional nature use. However, these are often only subjective assessments, since the problem of early diagnosis of diseases in the Arctic regions of Russia is particularly acute, and not only instruments and specialists are available in the district centers that could conduct regular medical checkups of the population but even a morgue, i.e., in rural settlements there is no way to establish reliably the cause of death. In most cases, early and sudden deaths, the local population refers to oncological diseases as the consequences of the activity of an industrial enterprise in the territory of their living. In the course of a poll among the inhabitants of indigenous communities, it was found that a high mortality rate is also associated with the problem of alcoholism and crimes committed under the influence of alcohol. The traditional types of economic activity associated with hunting and fishing also endanger life: water safety rules are not followed as well as dealing with weapons.

Among environmental problems, the majority of respondents noted the decline in traditional craft facilities, which is directly attributed to climate change (e.g., the wild reindeer changes its migration routes under the influence of this factor and, as in the case of the Republic of Sakha (Yakutia), goes to the Krasnoyarsk Territory). According to observations of indigenous peoples, winters become warmer, which is expressed in heavy snowfalls and increased winter temperatures. This leads to river spill in spring, flooding of villages, and loss of the fishing opportunity in the traditional way, as the fish goes deeper. Flood threatens another serious problem for traditional craft—broken trees, which the river carries, can break the seines, which means that an indigenous individual and his family can be deprived of food. Many of the representatives of indigenous communities also note the man-made factor—pollution of rivers due to the implementation of industrial projects, shipping, etc.

The Ust-Yanskiy region, the second researched area, has specificity concluded in a huge accumulated damage due to a previous gold extractive mine Kular and closed settlements (Vlasovo, Severniy) caused by mass outflow migration since 1998 when this mine was closed. The barbaric way of extracting gold from only the large and medium fractions, the pursuit of the indicators, led to the fact that there is still enough gold in the recycled dumps that can be produced. Since 2017, the license for processing and restoring Kular mine has been transferred to Arctic Capital LLC, which has undertaken the task of eliminating the accumulated environmental damage, recultivation of soil, employment of the local population among indigenous peoples in the newly discovered deposit, and procurement of traditional products (venison, fish, etc.). The concept of social responsibility of business comes to the Russian part of the Arctic, and it becomes one of the few ways to preserve indigenous community and people on the place of their original habitat (**Figure 8**).

Figure 8.
Accumulated environmental damage in Vlasovo, Ust-Yanskiy region, Yakutia (photo: V. Gassiy).

The specificity of the researched territory is its inaccessibility, which has a negative impact on the development of traditional spheres of economic activity. Producing objects of traditional nature use (fish, berries, furs) involves not only consumption for personal purposes but also the need to transport them to the market in larger settlements. The lack of roads and the high cost of transportation by air or auto trucks make economic activity (trade) by-products of traditional nature use almost impossible. In the Ust-Yanskiy area, the main source of income is the extraction of the mammoth tusk, which brings a significant income to the tribal communities and individual entrepreneurs. However, this type of activity requires special training (traditional knowledge, physical form, etc.) and technical equipment (pumps, boats, etc.). Although there are widespread cases of attempts by local residents to obtain tusks and without the necessary equipment, which leads to lethal incidents. On average, according to local residents, the "washing" season is about 100 days, for which one well-trained person can collect from 500 to 800 kg of tusks. In monetary terms, such a "crop" can fluctuate from 10 to 15 million rubles or 160,000–230,000 US$. Moreover, a hot summer with anomalous temperatures is considered by local hunters for tusks as a blessing, since actively melting permafrost itself gives away the hidden remains of ancient animals hidden for thousands of years. It should be noted that in the villages where the main activity is the extraction of the mammoth tusk, one can see expensive modern machinery that local authorities do not always have (**Figure 9**).

The purchased transport equipment allows local residents to develop trade between settlements within the region. Given their remoteness from each other, and the impossibility of year-round traffic, this is an important factor in actually helping people survive in such a harsh terrain. This fact makes indigenous peoples to adapt to the climate change in tundra in a unique way. For example, it is often possible to meet indigenous peoples who are using a winter mode of transport during the summer period, since flooded areas of the tundra do not allow movement on motorized wheeled vehicles, **Figure 10**.

Figure 10 shows a group of Evens moving on a snowmobile to their native village. In their opinion, in recent years the climate in the tundra has changed

Figure 9.
Type of transport vehicle in a Kazachye indigenous community, Ust-Yanskiy region (photo: V. Gassiy).

Figure 10.
Snowmobile in summer tundra on the way to Khayyr (even community), Ust-Yanskiy region, Yakutia (photo: V. Gassiy).

considerably: "Winters have become warmer, and summer is unstable: there can be both hot days and cold months when berries do not have time to ripen" (reindeer herder Nikolai, 43); "The deer goes North and does not come here because of the midges, which is very much due to the heat" (hunter Michael, 52) (**Figure 11**).

As a survey of the indigenous community showed, the traditional economy for the majority of local population ceases to be the basic criteria for determining the ethnic characteristics of the people. The high level of unemployment among indigenous peoples of the North, including the Evenks, Evens, and Dolgans, is complicated by the peculiarities of the sectoral structure of employment and the

Figure 11.
Example of private household in Kazachye (even community), Ust-Yanskiy region, Yakutia (photo: V. Gassiy).

Figure 12.
Rainwater harvesting for personal consumption, Khayyr (even community), Ust-Yanskiy region, Yakutia (photo: V. Gassiy).

qualification and educational level of the economically active population. The succession of generations in the traditional sectors of the North is gradually disappearing. Young people, being witnesses to the everyday, problematic life of the older generation, are of the opinion that work in reindeer husbandry, hunting, and fishing is not prestigious and does not bring sufficient income to create the corresponding financial situation of the family. The studied living conditions of indigenous communities on the territories of traditional nature use testify to the low level of social, communal, transport infrastructure development, which affects the behavior of the younger generation, their desire to go to the city or find work in extractive companies. "The benefits of civilization" in the form of the Internet, social web sources, and public amenities, along with climate changes, form challenges to the traditional way of life, undermining the age-old foundations of tribal communities. The domestic problems of indigenous peoples are one of the main reasons for the reluctance to remain on their land, to lead a traditional way of life, especially nomadic. Often villages in the territories of traditional residence are not provided with drinking water, and the only sources are river, rain water, or snow (**Figure 12**).

3. Conclusion

Thus, climate change in the Arctic for indigenous communities is not a prospect of the future, but a real threat to the traditional way of life, food security, and their habitat. We believe that ensuring the social status, decent level, and quality of life of the indigenous communities depends on the ways of preserving and developing the traditional economy on a new material, technical, and technological basis. Market relations in reindeer husbandry and hunting are constrained by the peculiarities of the nomadic way of life and the mentality of indigenous peoples. The theory and practice of managing changes in the territories of the traditional nature use of the Arctic require a critical rethinking of established views. In the coming years, new management approaches will be needed to quickly respond to changes in the Arctic territories, as climate change and global warming lead to the biggest social problem—changing the traditional way of life of indigenous peoples. On the other hand, industrial development expands the area of its presence in the Arctic, which creates not only challenges for the indigenous population but also the opportunities to preserve their culture, traditions, and crafts. In this regard, it is necessary to introduce into the practice of public administration the decision-making model for choosing investment projects based on the priorities of local development, the interests not only of the state and business but also of the indigenous communities [18]. Therefore, in order to solve the problem of survival and adaptation of Arctic indigenous communities in the context of climate change, a proactive reaction of science and practice is needed, based on complex sociological, ethno-cultural, ecological-economic, and statistical studies of traditional nature-use territories.

Author details

Violetta Gassiy
Action Group on Indigenous Involvement, International Arctic Science Committee
(IASC), Kuban State University, Russia

*Address all correspondence to: vgassiy@mail.ru

IntechOpen

References

[1] Decree of the President of the Russian Federation of April 19, 2017 No. 176 "On the Strategy of Ecological Safety of the Russian Federation for the Period Until 2025". http://www.kremlin.ru/acts/bank/41879

[2] Decree of the Government of the Russian Federation of April 3, 2013 No. 512-r "On Approval of the State Program "Energy Efficiency and Development of Energy" for 2013–2020". http://government.ru/docs/1171/

[3] Kotlyakov VM. On causes and effects of current climate changes. Journal of Atmospheric and Solar-Terrestrial Physics. 2012;**21**:110-114

[4] Romero D, Corralb MS, Pereira ÂG. Climate-related displacements of coastal communities in the Arctic: Engaging traditional knowledge in adaptation strategies and policies. Environmental Science & Policy. 2018;**85**:90-100

[5] Indigenous Empowerment is Vital for Climate Action. UN Climate Change; August 09, 2017. https://unfccc.int/news/indigenous-empowerment-is-vital-for-climate-action

[6] Gassiy V, Potravny I. The assessment of the socio-economic damage of the indigenous peoples due to industrial development of Russian Arctic. Czech Polar Reports. 2017;**2**(7):257-270. DOI: 10.5817/CPR2017-2-25

[7] Bulletin. Population size and migration of the Russian Federation, Federal Service of State Statistics. http://www.gks.ru/wps/wcm/connect/rosstat_main/rosstat/ru/statistics/population/demography/

[8] Information on hunting for 2016–2017. Federal Service of State Statistics. www.gks.ru/free_doc/doc_2016/bul_dr/ohrana/ohota16.xls

[9] The Arctic Zone of the Russian Federation/Federal Service of State Statistics. http://www.gks.ru/free_doc/new_site/region_stat/arc_zona.html

[10] Novoselov A, Potravnii I, Novoselova I, Gassiy V. Conflicts management in natural resources use and environment protection on the regional level. Journal of Environmental Management and Tourism. 2016;**3**(15):407-415

[11] Bogoyavlenskiy VI, Sizov OS, Bogoyavlenskiy IV. Remote detection of sites surface appearance gas and gas emissions in the Arctic: The Yamal peninsula. The Arctic: Ecology and Economy. 2016;**3**(23):4-15. http://www.ibrae.ac.ru/docs/3(23)2016_%D0%90%D1%80%D0%BA%D1%82%D0%B8%D0%BA%D0%B0/004_015_%20ARCTICA%203(23)%2009%202016.pdf

[12] Atle S. More than 800,000 reindeer vaccinated against anthrax. The Barents Observer. August 08, 2017. https://thebarentsobserver.com/en/life-and-public/2017/08/more-800000-reindeer-vaccinated-against-anthrax

[13] Hotez PJ. Neglected infections of poverty among the indigenous peoples of the Arctic. PLoS Neglected Tropical Diseases. 2010;**4**(1):e606. DOI: 10.1371/journal.pntd.0000606

[14] On the Detection of Dangerous Toxic Chemical Elements in Deer By-Products. Federal Service for Veterinary and Phytosanitary Surveillance. http://www.fsvps.ru/fsvps/news/ld/216626.html

[15] Makarov DA, Komarov AA, Ovcharenko VV, Nebera EA, et al. The study of the content of dioxins and toxic elements in the by-products and meat of the reindeer from 8 regions of Russia and the assessment of the harm to health

when using venison for food. Federal Service for Veterinary and Phytosanitary Surveillance: Official Report 2014–2016. http://www.vgnki.ru/assets/files/issledovaniya-oleni.pdf

[16] Vasilyeva AP. Influence of global warming on the welfare of the indigenous peoples living on the territory of the Sakha Republic (Yakutia). Almanac of Modern Science and Education. 2010;**10**(41):74-75

[17] The Impact of Global Climate Change on the Health of the Population of the Russian Arctic. UN Mission in the Russian Federation. 2008. http://www.unrussia.ru/doc/Arctic-ru.pdf

[18] Novoselov A, Potravny I, Novoselova I, Gassiy V. Selection of priority investment projects for the development of the Russian Arctic. Polar Science. 2017;**14**:68-77

www.ingramcontent.com/pod-product-compliance
Lightning Source LLC
Chambersburg PA
CBHW081232190326
41458CB00016B/5749